The Flip Side of Evolution

The Flip Side of Evolution

Kirk Stephenson

Rhaeadr
Publishing

First Printing: August 2011

ISBN 978-0-9837546-0-2

To Dylan, Ella, and Cade

Through good times and bad,
may you always know you are loved.

Contact Information

Author
Dr. Kirk Stephenson
kirk@theflipsideofevolution.com

Publisher
Rhaeadr Publishing
info@rhaeadr.org

The Flip Side of Evolution
www.theflipsideofevolution.com

Notices

CONTENTS

INTRODUCTION

Something Isn't Quite Right

As far back as high school I had a nagging feeling something wasn't quite right with the theory of evolution. I couldn't explain it, but something felt amiss. Other areas of science seemed to be backed by hard data; evolution by soft opinions. Charmin soft. Oh well, having more important things to worry about (most notably my high school sweetheart and Friday Night football), I swallowed the tale hook, line, and sinker.

Indeed, it would be difficult to read any biology textbook from a high school or major university and conclude that evolution was anything other than fact. This is understandable, as information that teaches away from evolution is filtered out. There is a side of evolution that is highly touted for public consumption, and there is a flip side of evolution that is whispered behind closed doors.

Like most college students, I was taught the merits of evolutionary concepts such as the Big Bang, the formation of life from nonliving matter, and the evolution of complex species from simple species ("descent with modification"). And like most college students, I graduated with the impression that evolution should be viewed as fact—cold, hard scientific fact. To deny evolution would be tantamount to denying gravity. Any red-hot doubts I once held had been reduced to a few flickering embers.

And why not? Teachers and professors I liked and highly respected made compelling arguments, and facts to the contrary never fell on my ears. Evolution appeared to be an open-and-shut case. If it were arguable, why didn't I hear any arguments? I heard more debate as to whether or not chili should have beans (it should not) than on the scientific merits of evolution.

Then came graduate school where, as a student of chemistry, I was encouraged to think critically—without exception to accepted norms. Professors teach graduate students to challenge or "push back" on scientific claims. Many times I was forced to prove, modify, or abandon my scientific claims. No offense was taken; that's the way science works. Scientists advance the pool of knowledge as they run experiments and gather data to defend their points of view, or to challenge the views of others.

A View Without Blinders

After graduate school, I would occasionally pass the topic of evolution. My old acquaintance seemed different. Then again, maybe it was just me. After all, my vision had changed—I now looked through the eyes of an objective scientist rather than gullible spectator. My peripheral vision was also much improved, as if blinders I never knew existed had been removed. Old, latent suspicions were rekindled. I decided to revisit what I had previously accepted as an impressionable student.

What I discovered amazed me. The theory falls apart at the most basic level. Almost every tenet of evolution is arguable. I found no evidence for the formation of life from the nonliving, precious little for the evolution of complex species from simpler ones, and an evolutionary mechanism that cannot account for complex biological systems.

Perhaps more amazing than the lack of scientific evidence for evolution is the noticeable absence of evidence against it. Evolutionists seem more like lawyers than scientists, intent on finding data to prove a particular view, showing little to no interest in data that argues to the contrary (other than trying to keep it out of the classrooms). I had to look outside the boundaries of the "scientific establishment" to find ideas and facts that don't fit well with the theory of evolution. I found them in spades.

Despite the fact that most scientists are educated in institutions where evolution holds a monopoly position, many have serious reservations about, or outright reject, the theory on scientific grounds. What do these scientists know that provides them a sound basis for rejecting evolution as "fact"? They are aware of caches of compelling scientific information that run to the contrary. Fortunately, many who have evaluated the theory of evolution and found it wanting have recorded their opinions and findings for others.

A Trek Thru Time

This book, essentially an introduction to that information, provides a glimpse into the evidence and logic that lead a growing number of scientists to doubt the validity of evolution. What follows is a trek through history aligned loosely in chronology with evolution. It begins with the formation of the universe from a speck of material smaller than the size of a magic bean, and ends with the triumphant rise of man

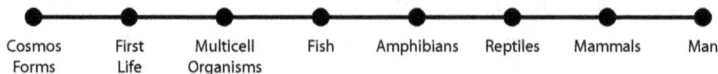

| Cosmos Forms | First Life | Multicell Organisms | Fish | Amphibians | Reptiles | Mammals | Man |

twenty billion years later. The trip is filled with many side excursions into basic scientific concepts necessary for background information and context.

The book is divided into five parts, each part consisting of two or more short chapters.

- PART I: In the Beginning
- PART II: Life from the Nonliving
- PART III: Simple Life
- PART IV: Fish to Philosopher
- PART V: The Long Journey Done

These sections stand alone, and may be read in any order. Part II is a little heavier reading than the other parts.

My purpose is to introduce the reader to facts and anecdotes about evolution that fail to get fair play in educational publications, and to provide a background for further study on the subject. After you have considered the information, I encourage you to determine for yourself whether the material would have broadened your perspective and enhanced your learning experience as a student.

If you disagree with what is presented, that's O.K. You are allowed to push back. The most important thing is that you do so in light of *all* the evidence, including evidence on the flip side of evolution.

PART ONE

In the Beginning

~ 1 ~

THE BIG BANG

Something From Nothing Equals Everything

My daughter, a drama major, was required to take a biology class in college. Although she failed to fully grasp the link between biology and stage acting, she managed to survive the ordeal, and I wound up with her 7-1/2-pound textbook entitled, appropriately enough, *Biology*. In that book, evolution is credited as "the greatest unifying concept of biology."[1] Quite an understatement. Evolution has done a better job of unifying biologists than a long chain with shackles does a crew of prisoners.

The theory of evolution provides the foundation for modern biology classes in college, high school, and even lower grades. This theory is so ingrained in our culture that the "theory" part is usually dropped. Most biologists today accept evolution as fact.

In essence, evolution teaches that all plants and animals evolved from a single, one-cell organism. Evolution's success is based on the marriage of two mechanisms: (1) chance *variations*, which come about in the form of mutations, and (2) *natural selection*, better known as "survival of the fittest."

In a nutshell, here's how it's supposed to work. Mutated creatures are occasionally formed that are better fit to the environment than their ancestors. These newfound advantages allow the mutants to survive over competitors until they, too, are eventually replaced by even more advantaged offspring. The process of incremental improvement and species genocide—producing more fit species while killing off less fit species—repeats over millions of years to produce increasingly complex and increasingly fit organisms.

The current champion of nature's King-of-the-Hill game is the most complex animal of all: man. Many of the losers, or what's left of them, are buried in the fossil record.

The Mighty Speck

The inquiring mind may wonder where the original one-celled organism came from. According to evolution, it came about by the chance combination of nonliving molecules floating about in a primeval swamp. So where did the nonliving molecules come from? Or for that matter, where did the swamp come from?

No one really knows. Scientists put forth speculative theories, but none lends itself to verification by the scientific process. Still, explanations of the origins of the universe are sometimes provided in biology textbooks. The most popular explanation goes something like this.

In the beginning there was nothing at all: no space, no matter, no time. Then, about twenty billion years ago, a little speck appeared. The speck was tiny—much smaller than the head of a pin—but it was mighty. It contained all the material from which every star, planet, and moon (and a host of other intergalactic bric-a-brac) in the universe would eventually form. This little speck, extremely hot and compact, exploded in a big bang. The arrival and explosion of the little speck marked the birth of space, time, and matter. Hot gasses shot into space in every direction. Eventually they collected and condensed to form everything in the cosmos as we now know it, including our solar system.

The little speck that could—apparently did! Indeed, a case could be made that history is essentially the story of this mighty speck: where it came from, and how it gave rise to all planets, plants, and people.

The formation of the original speck, however, is burdened by at least two major concerns: it is unsubstantiated by scientific testing, and it violates the First Law of Thermodynamics.

The First Law of Thermodynamics

Scientific laws are often based on complex mathematical expressions. Fortunately, those expressions typically distil down to practical ideas and solutions that are easy to grasp. Albert Einstein conducted countless hours of serious cogitation to derive the familiar $E = mc^2$ (energy equals mass times the speed of light squared). Most persons without a degree in science can readily comprehend the gist of Einstein's equation. The average high school student can substitute for any two of the variables and correctly solve for the third. Working with the equation doesn't take a rocket scientist (even though developing it did).

The First Law of Thermodynamics may seem complicated, but like Einstein's law, it distills to a beautiful and simple concept. Basically this law states that something cannot form from nothing. In more scientific terms, the sum total of mass plus energy in the universe remains constant. It is today what it always was, and always will be.

Mass		Mass		Mass
+		+		+
Energy	=	Energy	=	Energy
PAST		*PRESENT*		*FUTURE*

Here's the clincher. If there was nothing to begin with, and if something cannot evolve from nothing, from whence did the original speck come? Here we see that (cosmic) evolution is at odds with the First Law of Thermodynamics because it necessitates the formation of every thing (originally in the form of a mighty speck) from nothing.

The First Law works a lot like a debit card. My card will not allow me to buy something from a bank account with nothing in it; and the First Law will not allow Mother Nature to create something from a universe with nothing in it. Try either, and the same response is returned, "Sorry, transaction not allowed."

~ 2 ~

SOMETHING FROM NOTHING

Beam Me Up, Scotty

Solving the paradox of the beginning of the universe as it relates to the current laws of science has the earmark of a tough day at the office. Laws of science as we know them do not allow for the formation of the original speck from nothing, yet the Big Bang demands that it form.

Given no way out of this real-world dilemma, some scientists retreat to the safe haven of metaphysics. Ahhh, that wonderful world of speculation unaccountable to the scientific process. Metaphysics: "A priori speculation upon questions that are unanswerable to scientific observation, analysis, or experiment."[2] If a beginning conflicts with the First Law of Thermodynamics, why not just do away with the beginning? Why not start with, say, a beginning-less beginning?

Explanations to abstract questions often rely on mathematical models. However, when mathematical models that cannot be proven are used to validate theories that cannot be tested, assumptions get piled upon assumptions. The joke has been made that Ph.D. is an abbreviation for "Pile it Higher and Deeper." The joke is not totally groundless.

To demonstrate how a mathematical expression can be used to rationalize a theoretical concept, I will use the illustrative equation:

$$0 = 1 + (-1)$$

We know that zero represents *nothing*, and we know that the number one is *something*. Hence, from *nothing* (the zero on the left-hand side of the equation) we can get a *something* (the one on the right-hand side of the equation). Presto! Something from nothing.

There remains, however, the ever-so-slight matter of the negative one (-1) on the right-hand side of the equation, which I will call a *something else*. Rewriting the above equation in words:

nothing = something + something else

This *something else* should be problematic, as no one has ever seen one or observed one forming. No one knows what it is. This is because it only exists in mathematical terms—it is a mathematical construct. It doesn't represent anything in the real world. It is, for all practical purposes, a fudge factor. In the words of a disappointed British child who, having scooped out the last bit of ice cream from her doubly thick-walled, bubble-bottomed glass and discovered it held about half the amount she anticipated, exclaimed, "It's a cheat."

If the zero in the above equation denotes the pre-Big Bang state in which there was nothing, and the number one the post-Big Bang universe in which we live, then the negative one represents what? Perpetual anti-matter? A perfectly counter-balancing negative force or mass of some sort? A parallel universe running in imaginary time?

Mathematics notwithstanding, the whole something-from-nothing thing resembles Star Trek fiction more than fact. Beam me up, Scotty.

~ 3 ~

FALSIFICATION

The Timex Test

The idea of "proving" a scientific concept is technically a misnomer. Granted, we all understand what is meant by the phrase, but it is a hair we in science like to split.

In practice, a theory stands until it is proven false. This means that to be scientific, a theory must offer the possibility to show, through experiment or observation, that it is false, if indeed it is. In other words, a theory must be *falsifiable*. Without falsifiability, a theory is not really scientific—it is mere conjecture.

The idea that evolution is not falsifiable has led some to argue that evolution is not a valid *scientific* theory. Want to disprove Newton? Drop an apple and measure its acceleration due to gravity. Want to disprove Einstein? Measure for curvature of light as it passes an object of large mass. Want to disprove Darwin? Too bad. Empirical experiments that offer the possibility of falsifying evolution are not to be found.

(Granted, part of the problem might be the total lack of effort by biologists to design and execute such experiments.)

To be legitimate a theory must pass the Timex Test: it must take a scientific lickin' and keep on tickin'. In this regard, Darwin's theory fails miserably. Evolution is a heavyweight boxer with a glass jaw who never takes a lick because opponents are not allowed to enter the ring. Afraid of outside competition, evolution will only box with itself. It is, however, a great shadow boxer. Classrooms are allowed to openly debate the details of *how* evolution occurs, but not *whether* it occurs.

> *Scientific methodology exists whenever theories are*
> *subjected to rigorous empirical testing, and it is*
> *absent wherever the practice is to protect a theory*
> *rather than to test it.*[3]
> —PHILLIP JOHNSON, *DARWIN ON TRIAL*

Falsification issues aside, experiments designed to substantiate evolution are underwhelming at best. An experiment that produces ordered structures from a random explosion (*a la* Big Bang) is non-existent; nothing remotely similar to the simplest self-replicating organism has been made from basic materials; laboratory experiments have failed to produce more complex creatures from less complex ones; and a mechanism that rationalizes the formation of increasingly complex body parts or genetic information by gradual changes is sorely lacking.

Without laboratory experiments that mimic evolution, Darwinists rely largely on subjective interpretations of historical data (the fossil record) and comparative morphology (comparison of analogous body parts across species). The good news for evolutionists: because the data is subjective, they can "spin" or "hand wave" it to fit their theory. The bad news: because it is subjective, others can use the same data to argue against evolution. For example, evolutionists see in the fossil record a relatively smooth trend over time from less complex to more complex organisms, and conclude all organisms descended from a simple common ancestor; whereas anti-evolutionists see a gappy fossil record in which organisms appear fully formed and fully functional from their onset, with "missing links" between all major groups of organisms, and conclude the claim of common ancestry is false. Such is the nature of subjective data: one man's poison is another man's medicine.

The fact that evolution does not readily lend itself to the scientific process of falsification turns out to be a double-edged sword: evolution enjoys popular support because it cannot be proven false, yet it remains suspect for the same reason.

PART TWO

Life from the Nonliving

~ 4 ~

LIFE BEGINS ON EARTH

Hordes of Typing Monkeys

Early on, earth was a hot orb of liquid metals and other materials traceable to the Big Bang. The earth's surface, so the story continues, cooled to the point that an outer shell or crust formed. It was strewn with primordial pools of various elements and small molecules, many of which collided to form yet larger molecules through the process of *chemical evolution*. Molecules continued to collide and react until the pools contained high concentrations of large and biologically important molecules such as nucleic acids, proteins, and polysaccharides.

Then it happened. Mother Nature, a mere girl at the time, somehow arranged things so that the right kinds of large molecules collided at exactly the right time, at the correct angles, and with sufficient energy to produce a living thing. Eureka! Abiogenesis—life from the nonliving!

Were it to have actually occurred, the creation of a living entity from inanimate materials would be the most spectacular event since the Big Bang. The simplest form of life is extremely complex—and that's a gross understatement. The space shuttle, with its maze of electronic gadgetry connected by 235 miles of wire, is mere child's play by comparison.

To describe any life form as "simple" is to do it a great injustice. To sustain life, the very first organism would have needed a cargo load of highly integrated life-support systems, including extremely intricate processes to:

- take in energy
- convert energy into a usable form to run all biological processes
- eliminate waste products to prevent poisoning
- reproduce itself to propagate the species

To get a sense of just how complex our first self-replicating organism would likely have been, consider the bacterium *E. coli*. It is one of the simplest organisms living today. To provide itself with the systems it needs to live and replicate, this "simple" little one-cell wonder packs a lot of information in its DNA—tens of millions of bits, comparable to the amount of information in a couple of encyclopedias. With the information encoded in its 250 or so genes, *E. coli* builds molecular workshops, which build molecular machines, which undertake the myriad of complex functions necessary to maintain and reproduce life.

Without all complex processes and biochemicals in place and working in unison from the get-go, the first living organism dies without leaving offspring, and all is lost. No plants. No animals. No NCAA football.

Hoyle and The Typing Monkeys

What are the chances such an organism could have formed in one fell swoop by chance? Renowned astronomer and mathematician Sir Fred Hoyle (1915-2001) thought it is about as likely as a horde of monkeys accidentally typing, without error, the complete works of Shakespeare. He put the odds at less than one to a number with forty thousand zeros after it that life could form spontaneously from non-living materials.

Hoyle was correct. To understand why this is so, we need to introduce the concept of the *universal probability bound*.

The Universal Probability Bound

Although it seems infinite, the universe has a finite age and is composed of a finite number of particles that have limits on how fast they can interact. As a consequence, only a finite number of "events"—for example, particles such as atoms and molecules colliding in a chemical reaction—has occurred since the Big Bang (a belittling term Hoyle coined to describe the alleged event).

From estimates of the mass and age of the universe and interaction rates of its particles, scientists and mathematicians have estimated the total number of events since the Big Bang. Estimates fall in the range of the number one with 50 zeros after it (denoted 10^{50}) to one with 150 zeros after it (10^{150}). For simplicity sake, let's assume the number in the middle, 10^{100}, is correct. This means the maximum number of events that could have happened in our known universe since the Big Bang is 10^{100}.

While 10^{100} is far from infinity, it is nevertheless a large number. Consider the unabridged version, the number one with one hundred zeros after it:

10,000,000,000,000,000,000,000,000,000,
000,000,000,000,000,000,000,000,000,
000,000,000,000,000,000,000,000,000,
000,000,000,000,000,000.

The size of this number becomes evident when compared to more familiar ones: ten million is represented by 10^7 (the chances of winning Texas Lotto is about 1 in 10^7); one billion is 10^9; one trillion is 10^{12}; and one quadrillion is 10^{15}. The number 10^{100} actually has its own name: a *googol*. (Yes, the web search engine Google is a play on the word googol.) The term googol is easier to remember than its official name of 10 duotrigintillion, which is easier to remember than 10 billion quadrillion, quadrillion, quadrillion, quadrillion, quadrillion, quadrillion.

A googol events is indeed a very large number of events, but is it large enough to give evolution a reasonable chance of forming, by blind luck, life from nonliving materials? Once you understand the basic concept of universal probability bound, the answer is straightforward. One merely needs to determine the chances for the occurrence of some target event (for example, synthesizing and integrating all the molecules necessary to make the simplest living organism), then compare it to the total number of chances (events) available since the beginning of our universe. If the chances of a targeted event having occurred are about 1 in 10^{100}, it could have happened—maybe once since the beginning of time. If the chances are much greater than 1 in 10^{100}, the possibility that it occurred becomes proportionately greater. But if the chances of a targeted event having occurred are considerably less than 1 in 10^{100} then it, in all likelihood, didn't happen.

Given its importance, the probability one-in-the-total-number-of-events-to-have-ever-occurred is given a special name: *the universal probability bound*. William Dembski, who coined the phrase, uses the conservative value $1/10^{150}$ (shorthand for 1 in 10^{150}) rather than our assumed value of $1/10^{100}$ (1 in a googol). He defines the universal probability bound as:

> A degree of improbability below which a specified event of that probability cannot reasonably be attributed to chance regardless of whatever probabilitistic resources from the known universe are factored in.[4]

In short, the chance is essentially zero that any specified event with odds less than the universal probability bound will occur.

Let's make full circle and return to Hoyle, who set the odds for chance formation of life from nonliving materials at 1 in $10^{40,000}$. Hoyle's probability is vanishingly small when compared to the universal probability bound. If you think odds of 1 in 10^{100} is small, try 1 in $10^{40,000}$! The unabridged version of Hoyle's number would require enough zeros to overfill this book.

The possibility that life formed from nonliving materials simply does not exist. There is too little time since the Big Bang, and too few events. You might as well try to roll a seven on a six-sided die. You can roll the poor thing until it looks like a marble and you will not get a seven. Abiogenesis never happened—it's not even a close call.

~ 5 ~

SMALL MOLECULES BY CHANCE

Cooking with Miller and Urey

Hoyle may have closed the front door on abiogenesis (formation of living entities from non-living materials), but he left the back door wide open. Realizing that the odds of abiogenesis occurring in one fell swoop are nil, some suggest an incremental approach. They propose constructing life one building block at a time, then assembling the building blocks to make the whole. The odds of forming life from prefabricated building blocks are better than forming life from scratch.

Chemical Evolution in a Nutshell

The incremental approach assumes that small molecules in the primordial swamp reacted to form molecules only slightly larger. These, in turn, reacted to form molecules slightly larger still. This process of chemical evolution continued until a myriad of very small increases eventually yielded the large building blocks necessary for life. It was then a relatively simple task to combine the large prefabricated macromolecules into a living entity.

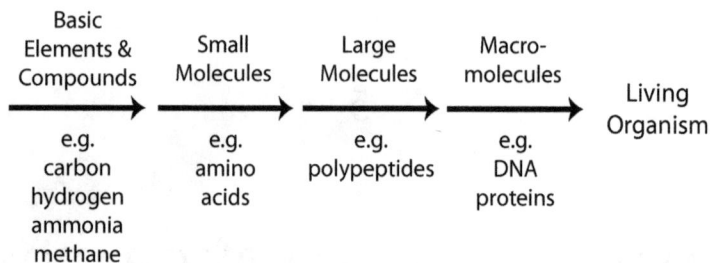

Basic Elements & Compounds	Small Molecules	Large Molecules	Macro-molecules	Living Organism
→	→	→	→	
e.g. carbon hydrogen ammonia methane	e.g. amino acids	e.g. polypeptides	e.g. DNA proteins	

The idea seems to be that the steepest of mountains is best traversed with the smallest of steps. It sounds a little like what Sen. Everett Dirksen is attributed as having said of the U.S. national debt: "A billion here, a billion there, and pretty soon you're talking real money." Maybe that's it: a large molecule here, a large molecule there, and pretty soon you're talking real life.

The many-small-steps scenario might go something like this. From the basic elements, nitrogen bases were formed. Ribose, a sugar, was also spontaneously formed. Ribose and the nitrogen bases reacted with phosphate groups to form nucleotides. The nucleotides combined to form nucleic acids, which combined to give RNA, and eventually DNA. Then Girl Nature, playing with her molecular Lego kit, used biological macro-structures such as ribosomes to snap together amino acids in the exact order specified by the DNA code, thereby forming specific proteins. Once these self-replicating chemical workshops started, like the Eveready Bunny they just kept going and going and going, generation after generation after generation, ultimately producing, among other things, the reader of this book—you!

The Miller–Urey Experiment

Years ago, when I was a gullible undergraduate student with a skull full of mush, my biology professor served up the story of Harold Urey and Stanley Miller. Their claim to fame sprang from experiments in which they virtually created life in a test tube, or so our class was led to believe.

Miller and Urey created their version of a primordial soup, composed primarily of methane, ammonia, hydrogen, and water. They placed these simple materials into a special apparatus from which all air had been removed. A continuous spark, which mimicked lightning, was used as a source of energy to facilitate the reaction of these simple materials.

In short, they gathered a bunch of nonliving materials, placed them in a carefully designed apparatus, and struck it with lightning. (For some reason, the movie *Frankenstein* comes to mind.) Amazingly, they made some amino acids, from which life-giving proteins are made. Abiogenesis basically proven. It's alive!

Forget the impression that something was created which barely missed having the ability to crawl out of the flask and bite you on the knuckles. The primary component formed in their experiment was something other than the stuff of life. About half of the amino acids used to build living things were produced in low yield, most in trace amounts.

What about other stuff necessary for life?

- How about nucleotides? No.
- Any proteins formed? None.
- How about other materials important for life, such as lipids? No.
- Sugars or carbohydrates? Not so much.
- DNA? RNA? Uh . . . no.

A Soup Pot or a Still?

The apparatus Miller and Urey used was not as basic as some are led to believe. Because biologists speak of a primeval soup, one might assume Miller and Urey used something like a soup pot. Actually, the carefully designed apparatus looked more like a still. Water was boiled in one area to produce water vapor. In another, the primeval environment (methane, ammonia, hydrogen, and water vapor) was exposed to low-amperage sparks. All oxygen in the simulated environment was carefully eliminated. After the reactions in the gas chamber, the gaseous products were condensed into liquids, and amino acids and other organic materials isolated from the mixture in a trap. On the whole, there seems to have been a lot of "design" and very little left to "chance."

The simulated primeval environment appears little more than a conveniently contrived reaction mixture formulated to yield preconceived materials. In other words, it was a targeted synthesis. The continuous low-amperage sparks mimic a long bout of lightning strikes at the same location, I suppose. The chemical trap? It's a mystery as to what in nature might have served as a viable chemical trap to selectively isolate specific amino acids from other components, and facilitate their conversion into proteins.

The Oxygen-Ultraviolet Conundrum

The Miller and Urey experiment is riddled with what should be fatal flaws, yet like Dracula this tale will not die. It is invulnerable to any silver bullet science has to offer.

Consider the issue of oxygen. Miller and Urey ran their experiments under a reducing environment, meaning that no oxygen was present. It was indeed fortunate that their hypothetical primitive environment just happened to be an oxygen-free ("reducing") environment, because oxygen stymies the very type of biological building blocks they were hoping to make.

However, it now appears that the primitive environment probably *did* contain oxygen.[5] Oops. This, among other things, reduces Miller and Urey's experiment to an irrelevant exercise in organic synthesis, as there is no dispute that amino acids can be synthesized in the laboratory under artificial conditions.

Regardless, the assumption of an oxygen-less primordial atmosphere might do more harm than good. No oxygen (O_2) means no ozone (O_3), an important compound that shields Earth from excessive amounts of ultraviolet (UV) radiation. Without ozone, UV radiation would have decomposed amino acids as they formed.

Michael Denton, in his book *Evolution: A Theory in Crisis*, summarizes the *Oxygen-Ultraviolet conundrum*: "What we have then is a sort of 'Catch 22' situation. If we have oxygen we have no organic compounds, but if we don't have oxygen we have none either."[6]

There are yet other problems, such as the lack of evidence in ancient rocks for the assumed primordial environment, but suffice it to say that it is a bit disappointing, though not terribly surprising, to find college textbooks void of anything that resembles a critical assessment of Miller and Urey's experiment. Worse still, it is hailed as one of the great icons of evolution. (However, in his book *Icons of Evolution*, Jonathan Wells lets most of the hot air out of the over-inflated Miller-Urey balloon.)

It Happens All the Time

The question is not whether chemists can synthesize many of the building blocks of life under carefully controlled laboratory conditions. Of course they can. There is an entire area of chemistry devoted to the subject of protein synthesis. Extremely complex syntheses are well within the capabilities of today's organic chemists. It's really no big deal. Pharmaceutical companies manufacture an amazing array of very complicated molecules, including "natural" products, all the time. Thyroid gland doesn't work? Not a problem. Using carefully designed experiments, chemists can make the same chemicals your thyroid gland is supposed to make. The real question is whether the building blocks of life can be produced (a) under conditions similar to those of the primeval soup and (b) without the guidance of some intelligent agent such as a chemist.[7]

The apparatus and reaction conditions of Miller and Urey, though carefully designed and controlled, mimic the real world about as well as a B52 stealth bomber does a toaster oven. Their experiment merely verifies what we already know: that knowledgeable chemists can synthesize basic chemicals under controlled laboratory conditions.

Wooden stake, anyone?

~ 6 ~

LARGE MOLECULES BY CHANCE

Much Ado About Nothing

For the sake of discussion, let's concede the implausible. Let's concede small molecules requisite to life somehow formed. Let's also concede their concentrations were high enough, and their proximity close enough, to allow for chemical reactions. Furthermore, assume there were no degenerative or unfavorable side reactions due to the presence (or absence) of oxygen, ultraviolet radiation, or anything else.

With the basic ingredients and favorable reaction conditions in place, could the thousands of macromolecules of life—such as ribosomes, RNA, DNA, lysomes, perisomes and other proteins—have been created by accident?

O.K., forget thousands. How about just one. What are the chances of making, say, one lousy protein by fluke?

Accidental Proteins: What Are the Chances?

Before considering the probabilities, it is necessary to provide a little chemistry background. Proteins are chemical chains made of amino acid links. Twenty types of amino acids are used to make living things (although the total number of amino acid types is in the hundreds).

To keep the math simple, let's choose to make a one-hundred-link protein. (An average protein is about four hundred amino acid links long.) Think of our simple protein as a one-hundred-link chain in which each link is one of twenty colors, each color representing one of the twenty amino acids.

In this special chain, each of the colors must be in the correct sequence for the chain to perform its function. The order in which the links is connected is critical. Get the color of even one link out of order, and bad things can happen. For example, when one amino acid (valine) is incorrectly substituted for another (glutamic acid) at only one site in the beta chain of hemoglobin, sickle-cell anemia occurs. (The hemoglobin beta chain contains 146 amino acids, with the incorrect substitution (mutation) occurring at link 6.) This single error causes blood cells to become sickle-shaped and rigid after they release oxygen, resulting in vessel blockage and blood flow complications.

Now, what is the probability of forming our perfect one-hundred-unit protein by sheer luck? Let's take it in steps. Because there are twenty amino acids from which to choose, the chances of randomly selecting the correct amino acid for the first of the one hundred positions are 1 in 20, or 1/20. The chances of putting the correct amino acid at the second position are also 1/20. Hence, the odds of placing the correct amino acids in the first two positions are

$$(1/20) \times (1/20) = 1/400$$

or 1 in 400. The odds of placing amino acids in the first three positions correctly are

$$(1/20) \times (1/20) \times (1/20) = 1/8000$$

or 1 in 8000. Continuing along the same line of reasoning, the chances of placing all one hundred positions correctly are found by multiplying 1/20 by itself for all 100 positions:

$$(1/20)_1 \times (1/20)_2 \ldots \times (1/20)_{100} = 1/10^{130}$$

The fraction 1/20 multiplied by itself for all one hundred positions equals approximately 1 in 10^{130}.

In trying to hazard one simple protein we are well below the zero probability threshold (universal probability bound) of 1 in 10^{100}!

For those who prefer to think in terms of letters and words rather than amino acids and proteins, let's play a modified version of *Wheel of Fortune*. Vanna White has before her one hundred blank blocks, each representing a letter, space, or punctuation mark.

◻◻◻◻◻◻◻◻◻◻◻◻◻◻◻◻◻◻◻◻
◻◻◻◻◻◻◻◻◻◻◻◻◻◻◻◻◻◻◻◻
◻◻◻◻◻◻◻◻◻◻◻◻◻◻◻◻◻◻◻◻
◻◻◻◻◻◻◻◻◻◻◻◻◻◻◻◻◻◻◻◻
◻◻◻◻◻◻◻◻◻◻◻◻◻◻◻◻◻◻◻◻

For each of the one hundred positions in the phrase (analogous to the one hundred link protein), there is a corresponding bucket containing 20 choices (analogous to the 20 amino acids): the correct choice plus nineteen incorrect ones.

Your job is to pull a character from each of the one hundred buckets, while blindfolded, and hand them to Vanna. She will place your random selections in the corresponding boxes. (This is essentially what Hoyle's monkeys were doing, only typing letters rather than drawing them from a bucket.)

Here you go. Mentally reach into the first bucket, select a character, and hand it to Vanna for placement in the first box. Your chances of drawing the correct character are, of course, 1 in 20. Now, move to the second bucket and select one of the 20 characters for placement into the second box. Continue until characters have been drawn from the remaining 98 buckets and placed in their respective boxes.

Now Vanna reveals the correct phrase. Did you get it right?

I can tell you with certainty that you did not. The chances the correct one-hundred-character sentence can be constructed in any one pass is 1 in 10^{130}, well below the universal probability bound and for all practical purposes, zero.

What If I Try a Lot of Times?

You can try as many times as you like and as often as you like, but you will not be successful. Only 10^{18} seconds have ticked away since the Big Bang. This means that if you somehow drew simultaneously from all one hundred buckets each and every second since the Big Bang, you would only have 1,000,000,000,000,000,000 tries at constructing the phrase. That would only increase your chances of success to about 1 in 10^{112} ($1/10^{130} \times 10^{18} = 1/10^{112}$), still well below any reasonable chance of success. Even twenty billion years isn't long enough for a simple protein to form by accident. In the end, all you will have is a jumbled sentence and very tired arms. The hopelessness of abiogenesis from random collisions of atoms and molecules is readily apparent.

What If I Target More Than One Protein?

Someone might reasonably protest the above calculations are for a *specific* protein, but the human body has as many as four million proteins. So, with four million targets rather than just one, aren't the chances improved?

Indeed they are—but not nearly enough. If we generously assume the first living cell had as many as ten million (10^7) different proteins, the chances of forming any one of them by accident are only 1 in 10^{123} ($1/10^{130} \times 10^7 = 1/10^{123}$). This number is one hundred thousand quadrillion times smaller than the universal probability bound.

The Rest of the Story: Self-Organization

Perchance all four thousand proteins and every other molecule requisite for life had been synthesized at the same time and location, there remains the insurmountable issue of self-organization. Mother Nature could no more accomplish such a feat by chance than you could shake a shoebox with watch parts and produce a working timepiece.

The organization and complexity of the smallest self-replicating cells is staggering. Its genome (total genetic code), probably in the ballpark of half a million base pairs (the human cell has about three billion base pairs), provides the information necessary to:

> ... code for proteins that are properly matched
> to their functions as part of the large cellular
> machines, such as the ribosome, which are re-
> quired even in the simplest of organisms. The
> expression control system must be able to pro-
> duce the proteins when they are needed, and in
> the amounts that are needed, in the presence
> of reagents needed to transform the proteins
> into active forms, and under conditions favor-
> ing proper folding of each protein.[8]

To further complicate the matter, buried in all this is a Catch-22 dilemma. Cellular workshops that assemble protein chains from amino acid links are themselves made from proteins. If it takes one to make one, from where did the first come?

As to Vanna's phrase above, it can be found in Shakespeare's *Much Ado About Nothing*:

> Don Pedro. Nay, if Cupid have not spent all
> his quiver in Venice, thou wilt quake for this
> shortly.

Given the improbability of constructing this nineteen-word sentence by chance, it is easy to appreciate Hoyle's belief that hordes of monkeys typing for eternity could not construct, and arrange in proper order, the more than 850,000 words in the complete works of Shakespeare. Infinitely less likely is the chance synthesis, and subsequent organization, of the set of compounds necessary to construct the simplest living cell.

~ 7 ~

CHIRALITY

Hand Cuffed

Actually, the chances of randomly forming a biologically active, one-hundred-unit protein are much less than previously indicated. The diminished odds are due to the fact that, for proteins to be biologically active, they must be left-handed.

Left-handed?

Handedness

That's right. Amino acids have what is called a chiral center. This forces them into one of two spatial ("3-D") configurations, which are referred to as right-handed or left-handed. Although right-handed and left-handed pairs of molecules have the same chemical makeup (i.e., identical formulas) they are not interchangeable because they are spatially different—their configurations differ in three-dimensional space.

Molecules that are non-superimposable on their mirror images are said to be *chiral*.

As a simple illustration, consider your left and right hands. They are like chiral molecules. Although they have the same components (fingers, thumbs, palms, backhand), they have different spatial arrangements. Imagine awakening one morning to find your left hand on your right arm and your right hand on your left arm. Freakish looks aside, you would not be able to function very well. Imagine the awkwardness in tying to steer a car, shoot a gun, or write a letter. And that softball glove you used last year—you might want to re-think that one.

Chirality ("handedness") at the molecular level is illustrated in the following diagram. Note that the molecules are mirror images.

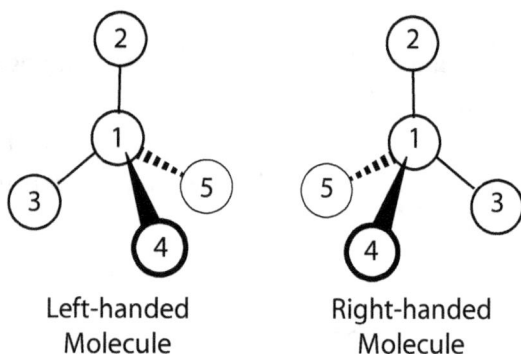

Left-handed
Molecule

Right-handed
Molecule

Consider the spatial arrangements. Each molecule is made of five atoms: atom 1, atom 2, atom 3, atom 4, and atom 5. Atom 1 is the chiral center because it has four different groups attached to it. Pretend atoms 1, 2, and 3 are in the plane of this paper (solid line bonds), the atom 4 is above the plane of this page (dark, wedge-shaped bond), and the atom 5 is below the plane of this book (dotted line bond). Within those constraints, mentally lift the molecules from the paper and try to superimpose them. That is to say, try to arrange them so that each atom on one molecule exactly coincides with the corresponding atom on the other. The molecules can no more be superimposed than your left and right hands. No matter how you overlay the molecules, at least two of atoms won't match up. The molecules are the same in chemical composition, but they are different in their arrangement in 3-dimensional space.

The difference may not mean much to you, but it does to Mother Nature.

The Miller-Urey Experiment Revisited

Mother Nature turns out to be very particular about the amino acids she uses to build proteins—she will only use the left-handed ones. A protein composed of both right- and left-handed amino acids would not have biological activity. It would be useless. Had Mother Nature accidentally mixed in right-handed amino acids when she built your proteins, you would never have lived to know it. Building a living entity is an exacting process.

Remember the well designed Miller and Urey experiment? It yielded a mixture of both left-handed and right-handed types of amino acids. Had the amino acids somehow managed to link together, the resulting protein would have been composed of both right- and left-handed acids, making it useless as a precursor to life. For biologically active proteins to form under Miller-Urey type conditions, left-handed amino acids would have to be cherry-picked from the primordial chemical brew or find each other by sheer luck.

The Odds: 100 Heads in a Row

Given evolution is a blind process that operates purely by happenstance, what are the odds that only left-handed amino acids would appear in a simple one-hundred-unit protein made in the primordial soup?

The statistics on this one is easy. At any point during the production of our one-hundred-unit protein, there is a 50-50 chance that a left-handed amino acid will be selected from our mixture. Statistically, then, the chances of selecting only left-handed molecules are the same as flipping a coin and it landing heads (50-50 chance) one hundred times in a row (length of our hypothetical protein). That probability is about 1 in 10^{30}, or 1 in a nonillian (a quadrillion quadrillion).

Earlier we learned that the odds of randomly arranging one hundred amino acids into a specific 100-unit protein sequence (from among four million choices) are 1 in 10^{124}. These odds are further reduced to 1 in 10^{154} ($1/10^{124} \times 1/10^{30} = 1/10^{154}$) because only left-handed amino acids are allowed in the protein. Once again, simple mathematics makes a case against the accidental formation of even one puny biologically active protein, let alone the hundreds of complex proteins required to make a living entity.

In the 1960's, Dean Kenyon coauthored a book entitled *Biochemical Predestination*. His popular theory advocated that the self-assembly of small chemicals (e.g. amino acids) into macromolecules (e.g. proteins) was essentially inevitable. It was bound to happen on its own. However, the ever-objective Kenyon realized his theory could not explain, for example, how the self-assembly of intricately folded proteins could occur in the absence of detailed instructions such as that provided by DNA. The pioneer of chemical evolution remarks on the field's current status, "We do not have the slightest chance of a chemical evolutionary origin for even the simplest of cells."[9]

~ 8 ~

DIRECTED PANSPERMIA

Mother Nature's Start-up Kit

Many scientists, including Nobel laureates, have come to the reasonable conclusion that it was impossible for life to have started by mere chance. They realize that life needed a kick-start of some sort.

Evolutionists need to get their theoretical foot wedged in science's doorway. If our entire universe can be rationalized from one speck of energy-packed matter smaller than a bean, then surely every plant and animal can be rationalized from one simple cell. The trick, then, is coming up with that very first cell.

Some scientists of note have proposed the idea of *directed panspermia*, a fantastic hypothesis sprinkled with just a touch of desperation. "Directed" denotes a chosen target, and "pansperm" refers to a life seed. Directed panspermia is the idea that some advanced civilization in outer space directed to earth the life seeds (micro-organisms capable of self-replicating) necessary to start biological evolution.[10]

Even if true, directed panspermia merely kicks the can down the road. How did the aliens who seeded Earth get started? Did other aliens from yet other planets seed their planet?

As far as life on Earth is concerned, directed panspermia appears to be Mother Nature's startup kit, courtesy of friends from a galaxy far, far away.

No return address.

Just add water.

~ 9 ~

THE AD HOMINEM ATTACK

If You Can't Win the Game, Win the Fight

There is a sports adage to the effect, "If you can't win the game, win the fight." Frustrated with their inability to defend a point of view, some attack *ad hominem*—they attack the opponent rather than the opponent's position. A scientist might jeer another, "If you don't believe in evolution, you're not a 'real' scientist." This has all the legitimacy of the insult, "Real men don't eat quiche." (O.K., maybe a bad example.) The latest sophomoric variation is to refer to anti-Darwinists as "anti-scientists."

Darwinism apologists have branded those who do not believe in their brand of evolution as everything from ignorant to stupid to misguided to downright wicked.

> *I'm all in for civil debate, so long as you agree with me.*
> —ANONYMOUS

That's not to say the favor isn't occasionally returned. Such Darwinists, boorishly intolerant of alternative viewpoints, have been dubbed "Darwiniacs," a synthesis of Darwin-maniac.

Science Gets Ugly When Sacred Norms Are Challenged

Evolution might be in vogue, but science is not a popularity contest. Scientific laws are not passed by majority vote of some Congress of Science: they are what they are, regardless of man's desire to make them otherwise.

And that's a good thing, because "scientific consensus" sometimes gets the big stuff wrong. Geosynclinal theory was superseded by plate tectonics as an explanation for the origination of mountains. Chemists eventually gave up on the widely-held idea of phlogiston as the material component of fire. The earth-is-the-center-of-the-universe (geocentric) model finally gave way to a better explanation of astronomical observations, the heliocentric model.

But dogma dies hard. Galileo, who championed the heliocentric model, was forced to spend the latter years of his life under house arrest for his views. Science gets ugly when sacred norms are challenged.

Fast forward to the present. The debate surrounding Darwinism is more serious than a bunch of geeky scientists hurling insults as if they were children in a Little Tikes

Turtle throwing sand. At the roots of the debate, which extend deeper than the deepest fossil, lie their world views and morality. William Dembski, mathematician and philosopher who takes spears daily for his compelling arguments against Darwinism, comments on the intensity:

> We now face a Darwinian thought police that, save for employing physical violence, is as insidious as any secret police at ensuring conformity and rooting out dissent.[11]

Michael Behe, another favored target of old-school scientists, remarked that graduate students who speak skeptically of evolution place their careers in harm's way; and he advises them not to do so.[12] Open-mindedness and critical thinking only go so far in today's university studies.

History's Honor Roll

So, if you have been so insulted, don't feel bad—you are in good company. Count yours among the names of many great scientists and engineers who believed in a *designed* universe, including:[13]

- Johannes Kepler, founder of physical astronomy.
- Blaise Pascal, founder of the science of hydrostatics.
- Robert Boyle, father of modern chemistry.
- Isaac Newton, among his numerous accomplishments, he discovered the law of universal gravitation.
- John Dalton, formulator of the gas law of partial pressures.

- Samuel Morse, inventor of the telegraph.

- Louis Pasteur, who established the germ theory of disease and the remedial processes of pasteurization.

- Lord Kelvin, who established the scale of absolute temperatures. He also helped establish thermodynamics as a formal scientific discipline and formulated the first and second laws in precise terminology.

- Bernhard Riemann, German mathematician who developed the concept of non-Euclidean geometries.

- John Fleming, father of modern electronics and developer of the first electron tube.

- George Washington Carver, agricultural chemist who developed over four hundred products from peanuts and sweet potatoes.

- Charles Stine, inventor and former Director of Research for the DuPont Chemical Company.

- Wernher von Braun, leading engineer on the development of the V-2 rocket.

Today's Honor Roll

The roll of scientists, mathematicians, engineers, and philosophers persecuted by the scientific intelligentsia can now be extended to include a group of progressive thinkers pioneering a new theory known as ID, or Intelligent Design. ID provides a sound basis for determining whether or not something is formed by design. Unfortunately for evolutionists, application of ID to biological systems indicates the presence of design, which implies the influence of a designer. This, of course, is anathema to Darwinists, who hold evolution is necessarily the product of *unguided* natural forces.

Nevertheless, the true champions of open-mindedness persist in their endeavors. Names to be added to the Honor Roll include:

- William A Dembski, mathematician, rationalized the elimination of chance through small probabilities and *specified complexity*. A pioneer whose ideas have shifted the scientific debate.

- Michael J. Behe, biochemist, brought the concept of *irreducible complexity* to the forefront of science, and Darwinian evolution to its knees.

- Philip E, Johnson, lawyer and early visionary of modern intelligent design.

- Michael Denton, molecular biologist and medical doctor, expounded scholarly critiques of evolution in the 1980's, paving the way for other intellectuals.

- Jonathan Wells, molecular and cell biologist, among other things, provided critical analyses of popular "icons of evolution" such as *Archaeopteryx*, homology in vertebrate limbs, and Darwin's finches.

- Guillermo Gonzalez, astrophysicist, pioneer of the *galactic habitable zone*, and concept that Earth is especially well suited for both life and discovery, consistent with a designed universe.

More names could be added, including the entire host of Society Fellows at The International Society for Complexity, Information, and Design (ISCID),[14] and those who contribute to books such as *Uncommon Dissent: Intellectuals Who Find Darwinism Unconvincing*. But be assured, were the list complete, there would not be an ignorant or stupid or misguided or wicked one among the lot.

THE SECOND LAW
(PART I OF IV):
ENTROPY

Law and Disorder

At this point in his adventure, the Mighty Speck has appeared out of nowhere, fragmented to form every bit of material in the universe, including a pool of primordial goo on the planet earth, replete with chemical building blocks necessary to life. It was a tough eighteen billion years' work, but everything is finally lined up for an easy finish. All that remains to do in the next two billion years is to get the primordial chemicals to self-organize and evolve into every organism known today.

What the Mighty Speck didn't know is that one of its greatest obstacles lay ahead in the form of another law of science. And it's another of the big ones: the Second Law of Thermodynamics.

While the Second Law can get hairy (werewolf level), it cannot be avoided. Critical arguments revolve around the

Second Law. Now for the good news: as with many other laws of science, the basic concepts are readily comprehendible by non-scientists.

The Second Law: What Is It?

The Second Law deals with the concept of *entropy*. Entropy is "a process of degradation or running down or a trend to disorder."[15] It teaches that, left to their own devices, things become more *dis*ordered (or less ordered)—they rot, rust, tear up, and wear out. It's no wonder entropy is often viewed in negative terms: as entropy increases, the quality of things decreases.

Your everyday life experiences validate that entropy (disorder) naturally increases. Your car breaks down; it doesn't improve or even stay the same. The paint on your house gets old, falls apart, and begins to look horrible. (Unfortunately, pretty much the same thing happens to people.)

The Second Law of Thermodynamics can also be thought about in terms of energy usefulness. Changes to a system will inevitably result in a loss of *usable* energy. This is why the perpetual motion machine Uncle Bob has in his attic doesn't work. Every time the machine is operated, some energy is wasted. Friction might cause heat, which is lost to the surrounding environment and hence no longer available to do work. Eventually, the contraption loses so much energy it cannot keep itself going. Either energy has to be pumped in from an outside source or the machine stops—in either case, it ceases to be a perpetual motion machine.

So, during any spontaneous (naturally occurring) change, a portion of the total energy is "wasted." My old college physical chemistry textbook likens entropy to a signpost. The textbook puts it this way:

> [Entropy] is the signpost for the direction of spontaneous change: we simply have to look for the direction of change which leads to the greater dispersal [less useful form] of the total energy.[16]

Where our everyday lives are concerned, entropy's signpost along nature's highway might read something like "This Way to Nature's Junkyard." Usable energy is running down, and everything else is tearing up.

In short, nature tends to change in the direction of increasing entropy, as evidenced by increases in disorder and losses of useful energy.

more entropy ≈ more **dis**order ≈ less usable energy

Abiogenesis Versus The Second Law

If things naturally tear up, wear out, and become less organized, how could elements and simple molecules floating about in a primordial goo have become so highly ordered they came to life? How is it that those early molecules were able to thwart the Second Law of Thermodynamics (natural disordering) to yield the ultimate and most complex form of organization in the universe: life?

To better understand the difficulty, let's look at entropy (disorder) in light of a simplified illustration. A master bricklayer carefully organized individual bricks into the shape of a house. For years it stands majestically in the sunlight. Unfortunately, time—decades, centuries, or millennia—takes its toll on the bricklayer's masterpiece. The house ends up as a pile of bricks. The carefully stacked bricks (low entropy) comprising the house decreased into a mound of unorganized bricks (high entropy), as one familiar with the Second Law anticipates. Entropy increased and the house decreased.

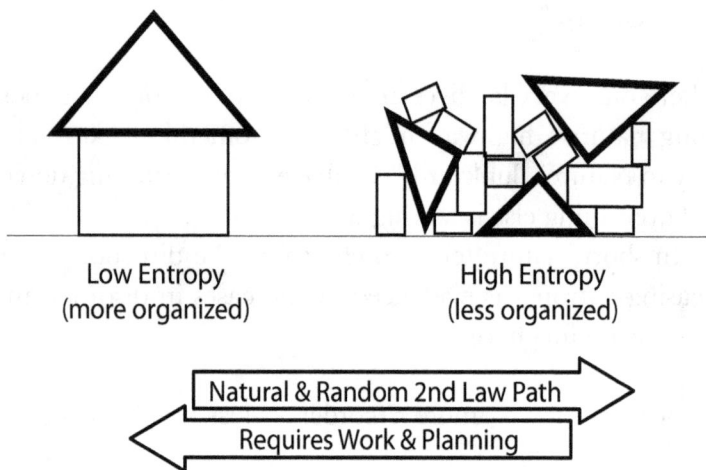

Low Entropy
(more organized)

High Entropy
(less organized)

Natural & Random 2nd Law Path
Requires Work & Planning

Of course, it *is* possible to build a house (increase complexity) by increasing the order of the bricks. But that's just it—it has to be built. It's not a natural, spontaneous process that occurs on its on. Given all the time in the world, the strewn bricks will not order themselves back into a house. Building a house is an unnatural process that requires orchestrated work (energy, effort). Bricks will not evolve into a house by chance, not in a million years . . . and not in twenty billion. Mother Nature will not stack the bricks for you, no matter how long you hold out.

In sharp contrast, evolution holds that a random assortment of chemicals somehow organized themselves into a highly complex one-cell creature by chance, a proposition that is infinitely more difficult than chance carpentry.

Evolution tells us that Nature can build a carpenter, but it cannot build a simple house.

~ 11 ~

THE SECOND LAW
(PART II OF IV):
THE OPEN SYSTEM ARGUMENT

A Closed Door

When a head-on challenge of the law does not work, clever lawyers resort to a tried and proven tactic: they look for a loophole. The same ploy may be applied to scientific laws. Where the Second Law is concerned, the hoped-for loophole comes in the form of the "open system" argument.

Entropy is typically defined in terms of a *closed* system. However, the primordial soup was actually an *open* system. Because the Second Law of Thermodynamics applies to closed systems whereas the primordial soup was an open system, some claim it is misapplied in the abiogenesis debate. Let me explain the argument, and then the fallacy.

What Are Open and Closed Systems?

In general terms, a closed system is one in which energy cannot enter from the outside. Because all changes within a closed system will disperse or waste some of its energy (as we learned in the previous chapter), the system will eventually run down. Wasted energy cannot be replenished in a closed system.

In contrast, an open system is one in which energy can enter the system from the outside. Energy dispersed (rendered useless) from activity inside the system can be replenished by the energy from the outside.

Closed System
(energy limited)

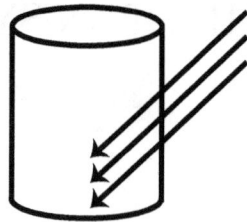

Open System
(extra energy available)

For simplicity, let's consider a closed system analogous to a locked room. If a basketball falls off a shelf onto the floor, each bounce will be lower than the preceding bounce because of energy dispersal (from friction, vibrations, air resistance, etc.). Eventually the ball will cease to bounce. For the ball to remain bouncing it needs to be acted on by some new source of energy—but it's a closed system (the door is locked) so none is available.

But what if energy is brought in from outside the system to replenish wasted energy? Could the system stay "alive" in such a case?

To see, let's turn our room into an open system by leaving the door ajar. This time, when our basketball falls off the shelf, someone rushes in and pushes down on the bouncing ball (supplying energy), thus keeping it bouncing. In a similar fashion, couldn't some external source have supplied enough energy to the primordial soup to keep the evolutionary ball bouncing? Couldn't energy from the outside more than compensate for the energy wasted inside the primeval system?

The Open System Argument and Abiogenesis

Some make the case that it could. The sun, they say, was the source of "outside" energy. It provided more than enough energy to compensate for any entropy decreases within the primeval pool.   Mother Nature used the extra energy from the sun to turn primordial chemicals into biomolecules, and

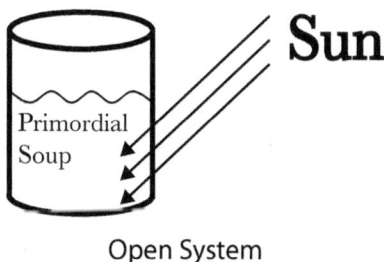

Open System

biomolecules into a living cell. The net result: energy from "outside the system" (the sun) made it possible to increase the entropy "inside the system" (the primordial soup). Energy from the sun apparently kept the evolutionary ball bouncing on earth.

Credit where credit is due: the open system argument does solve the entropy problem in a limited sense. During supposed abiogenesis, the entropy of the entire universe would have increased, consistent with the Second Law. The sun would have become less ordered (from the explosive chemical reactions it took to make the energy it sent to earth) than the primordial pool became ordered (from abiogenesis), hence a net increase in entropy in the universe as a whole. In essence, the price for local (primeval pool) ordering was paid by an increase in disorder somewhere else (the sun). Were this all there was to the story, evolutionists might have a point.

Problems with the Open System Argument

The most obvious problem with the open system argument is that it does not jibe with the real world. It is at odds with our everyday observations. The sun continuously provides energy to the earth and yet the Second Law continuously wreaks havoc. Things still rot, rust, tear up, and wear out. The same sun is shining now that shined back in the primeval days, and yet we see no indication of accidental increases in organization today. Why then and not now?

A larger and more technical problem relates to energy *quality*. Extra energy from the outside may be *necessary* for entropy to decrease in a locally defined system, but it is not *sufficient*. Returning to our brick house analogy, the sun can shine on our mound of bricks forever (thus providing more than sufficient *quantity* of energy) but it will not produce a house (because it was the wrong *quality* of energy). To order bricks into a house, the directed energy of a skilled bricklayer (high quality energy) is required—a mere excess of energy from the sun is insufficient.

While it is true that Darwin's primeval pool had more than enough energy to drive chemical reactions, it is also true that it lacked the right quality of energy to produce increases in complexity.

Four Requirements for Increasing Order and Complexity

Which brings us to the questions: Under what conditions can entropy be decreased? Besides excess energy, what else is required for local increases in order and complexity? The answer to those questions is the punch line to the entropy debate.

For complex ordering (simultaneous increase in both order and complexity) to occur, at least four things need be present:[17]

- An open system (opportunity to get energy)
- Available energy (necessary to do work)
- A coded plan (a blueprint, pattern, code, know-how)
- An energy-conversion mechanism (to apply energy in the "ordering" direction)

The pile of bricks bathing in sunlight will never become a house by chance because the last two items are missing. In contrast, a carpenter can build a house because, in addition to available energy (Items 1 & 2), he has a blueprint (Item 3) and arms for swinging the hammer in the correct direction (Item 4). Similarly, the basketball continues to bounce because the ballplayer brings Items 3 and 4 to the game.

The open system argument is not so much wrong as incomplete. An open system and outside energy (Items 1 and 2) would have been necessary for abiogenesis, but they would not have been sufficient. A fuselage is necessary for the construction of a passenger airplane, but it is not sufficient. Wings and a plethora of aeronautical gizmos are needed, too.

The primeval world was a world without a plan. It lacked Items 3 and 4. The sun can shine on a chemical cocktail from eternity past to eternity future and abiogenesis will never occur—not because there is not enough energy, but because there is no coded plan or mechanism to convert that excess energy into a usable (ordered) form.

Where abiogenesis is concerned, the open system argument is a closed door.

~ 12 ~

THE SECOND LAW
(PART III OF IV):
NATURAL INCREASES IN ORDER

How Is It That Gardens Grow?

If it's true that atoms and molecules can't spontaneously (naturally) order themselves into living things, how is it that seeds grow into plants? Mother Nature carefully extracts dead nutrients found indiscriminately in the soil, water, and atmosphere and stacks them neatly into living entities. If the Second Law of Thermodynamics does not allow for chemicals to become ordered by random acts of nature, how is it that gardens grow?

Fair question, and one that can be posed using other vehicles. If a bunch of molecules can be fashioned into a complex baby inside a mother, then why can't a bunch of molecules be fashioned into a one-cell organism inside a primordial pool?

Once again, the answer lies in the previously mentioned requirements for complex ordering:

- An open system (opportunity to get energy)
- Available energy (necessary to do work)
- A coded plan (a blueprint, pattern, code, know-how)
- An energy-conversion mechanism (to apply energy in the "ordering" direction)

In the development of a newborn baby, all four items are present. Items 1 and 2 are easy to come by; the mother has but to eat. Item 3 is inherent in the very first cell in the developing human. Its 100,000 genes, neatly folded into 23 pairs of chromosomes, contain the complex blueprint for assembling a human being. Each parent provides half of the DNA blueprint. Item 4 is provided courtesy of the mother, whose bio-machinery combines the two halves of genetic material, then uses the resultant DNA blueprint to arrange atoms and molecules into a child.

In contrast, a single-cell creature could not have formed in the primordial soup because Items 3 and 4 were missing. There were no parents with a preprogrammed code for making offspring. No genetic blueprint. Evolution does not have a coded plan for making the first living thing, and basic probabilities, as we have seen earlier, argue against it coming about by chance.

One of the great challenges for evolution is to provide a plausible explanation as to how the coded plans and energy-conversion mechanisms requisite for life came about in the first place, given that they couldn't have come about by a series of lucky accidents.

Growing a garden from seeds is easy; growing it from basic chemicals is impossible.

~ 13 ~

THE SECOND LAW
(PART IV OF IV):
CRYSTALLIZATION

Of Ice and Men

Fill an ice tray with water, place it in your freezer, and assuming your freezer is working, you will make ice. What's more, Mother Nature can do the same thing without the help of a Frigidaire: she can crystallize water into a snowflake or an iceberg on her own. No big deal, right?

As innocuous as they seem, snowflakes, icebergs and the like are used to argue that entropy can be decreased (organization increased) without the use of a coded plan or blueprint. Here's the rationale. Ice is more ordered (highly structured) than the water from which it is made. Ice is a crystalline structure of neatly stacked water molecules, whereas water in liquid form has relatively little structure. Thus, in making an iceberg from ocean water, Mother Nature increased the order of chemicals, the Second Law of Thermodynamics notwithstanding. Isn't it reasonable, then, to assume the same natural process that structures water into ice can also arrange amino acids into proteins, and macro-molecules into men?

Only if you are willing to overlook the concepts of molecular complexity, reversibility, and information content.

Stupid Is as Stupid Does

Darwinian evolution relies on the idea that complex molecules derive naturally from simple ones (chemical evolution). Merely cooling chemicals to the point they crystallize (freeze) into a more ordered state does nothing to increase molecular complexity. Ice and its less-structured precursor, water, are chemically identical. Both are H_2O. They differ only in physical state. In sharp contrast, complex organic macromolecules such as DNA and proteins are chemically distinct from their nucleic acid and amino acid precursors.

Furthermore, the process of forming ice from water is fully *reversible* (ice is easily converted back into water), whereas evolution is supposedly a one-way proposition.

Finally, there is the all-important issue of *information*. The key to evolution is ordering molecules with little or no information into ones with lots of information. For evolution to be successful, dumb nucleic acids must be converted to smart molecules like DNA, which contain all the information necessary to mold non-living materials into a living entity. Simple nucleic acids don't know how to do anything; DNA does. A bottle of blue ink splashed on a blank sheet of paper does not convey information; a blueprint does. In the case of crystallization there is no increase in information on going from liquid water to ice. Stupid is as stupid does.

In Summary

And with that, we come to the end of our discussion of the Second Law. As the last four chapters have shown, changes in physical state and sunshine do not remedy evolution's entropy problems.

Despite the availability of high-tech laboratories loaded with sophisticated equipment, modern biologists have made precious little progress in mimicking an event that supposedly happened in the great outdoors by happenstance. Left to chance, small molecules will not organize into biomolecules; and biomolecules will not organize into a living entity.

Darwinists need to somehow repeal or exempt Louis Pasteur (1822–1895) and Rudolf Virchow's (1821–1902) inconvenient Law of Biogenesis, which states that life only arises from previously existing life. Maybe a yet-to-be-discovered Law of Biological Ordering will arrive as a knight in shining armor to save abiogenesis from the evil entropy monster.

Until then, evolutionists take it on faith that nature somehow pulled the trick off.

If someone points out to you that your pet theory of the universe is in disagreement with Maxwell's equations—then so much the worse for Maxwell's equations. If it is found to be contradicted by observation—well, these experimentalists do bungle things sometimes. But if your theory is found to be against the second law of thermodynamics I can give you no hope; there is nothing for it but to collapse in deepest humiliation.

—ASTRONOMER SIR ARTHUR STANLEY EDDINGTON, *The Nature of the Physical World (1915)*

PART THREE
Simple Life

~ 14 ~

A (VERY) BRIEF HISTORY OF TIME

So Much To Do, So Little Time

Evolutionists claim time is the hero of the plot. Time performs miracles. Given enough time, anything is possible.

But how much is enough time? And what kinds of things have the miracles of time supposedly produced? To see, let's begin with a trip down evolution's memory lane.

History's Timeline

Twenty billion years ago the original speck of matter exploded in the Big Bang. Over the course of sixteen billion years or so, planet Earth formed from the resultant gases and subsequently cooled to the point that it became crusty. The earth was a pretty boring place—but that was about to change.

- *2,800 to 540 MYA (million years ago), Precambrian Period*
  One-cell organisms galore. Some multi-celled ancestors. Algae and bacteria. Lots of slime and ooze.

- *540 MYA to 500 MYA, Cambrian Period*
  "Age of Trilobites." Representatives of all major classes (phyla) of animals in existence today explode onto the scene, with no trace of progenitors. These include vertebrates and invertebrates, and marine animals with and without shells.

- *500 MYA to 440 MYA, Ordovician Period*
  Primitive plants appear on land. Primitive fish and seaweed appear in the oceans.

- *440 MYA to 410 MYA, Silurian Period*
  Jawed fishes, insects, and vascular plants show up. Centipedes and millipedes appear. (Intermediate 200- through 900-pede series failed to evolve for some reason.)

- *410 MYA to 360 MYA, Devonian Period*
  "Age of Fishes." Fish become abundant and diverse. Some of them begin walking on land in the form of amphibians. Land plants diversify.

- *360 MYA to 280 MYA, Carboniferous Period*
  Reptiles first appear. So do cockroaches and other things that go crunch in the night.

- *280 MYA to 250 MYA, Permian Period*
  "Age of Amphibians." Amphibians rule, but reptiles come on strong near the end. Those lovable trilobites, charter members of the Cambrian period, become extinct.

- *250 MYA to 210 MYA, Triassic Period*
  Reptiles get big—as in dinosaur big. Some become meat eaters—not good news for small mammals trying to get their species going.

- *210 MYA to 150 MYA, Jurassic Period*
  The original Jurassic Park. Dinosaurs galore. On the bright side, flowering plants start blooming, and birds show up.

- *150 MYA to 70 MYA, Cretaceous Period*
  Dinosaurs become extinct, but they leave behind a remnant in the form of alligators (most of which now homestead in Louisiana). Fruit begins to grow.

- *70 MYA to 1.8 MYA, Tertiary Period*
  With dinosaurs out of the way, mammals take over.

- *1.8 MYA to Today*
  Large mammals like the woolly mammoth are Ice Age casualties. Humans evolve. They become self-aware, figure out that they came from an explosion twenty billion years ago, and invent the iPad. Pinnacle reached.

To put the timeline in perspective, a crude geological timetable is shown on the next page but one. Time progresses from bottom to top. At the instant of the Big Bang, zero percent of our universe's twenty-billion-year history had transpired. By the time the Cambrian Period rolled around, about 97 percent of all history had already ticked away.

Time Is the Hero?

Multiple millions (if not billions) of favorable mutations would have been required to produce every variant of plant and animal alive today from a single, one-cell Precambrian organism, yet there are those who hold there was sufficient time for those mutations to have occurred. Evolutionist have held for decades that, given enough time, virtually anything is possible. Given two to four billion years, the unlikely becomes almost certain.[18]

Juxtapose the idea that all living organisms formed from scratch in four billion years against the astronomical improbability of a single one-hundred-unit protein forming by chance in twenty billion years, and shocking reality sets in: Time is not the hero of the plot—it is the villain!

PERIOD	*NOTE*	*MYA*	*PERCENT HISTORY SINCE THE BIG BANG*
Recorded history		0.006	present
Quaternary		1.8	99.9%
Tertiary		70	99.7%
Cretaceous	Age of Reptiles	150	99.3%
Jurassic	Age of Reptiles	210	99.0%
Triassic	Age of Reptiles	250	98.8%
Permian	Age of Amphibians	280	98.6%
Carboniferous		360	98.2%
Devonian	Age of Fishes	410	98.0%
Silurian		440	97.8%
Ordovician		500	97.5%
Cambrian	Age of Tribolites	540	97.3%
Precambium		2,560	87.2%
supercontinent		3,900	80.5%
crust cooling		4,600	77.0%
Big Bang		20,000	0.0%

MYA = millions of years ago

~ 15 ~

THE CAMBRIAN EXPLOSION

Biology's Big Bang Goes Bust

I used to believe the fossil record screamed evolution must be true. I have come to believe just the opposite, in large part due to the Cambrian period record. The discrepancy between Cambrian fossil data and evolution theory is so glaring that biologists give it a special name: *The Cambrian Explosion*.

According to evolution theory, the first one-cell creature arrived on the scene sometime in the Precambrian period. It was the sole member of what was at the time the entire plant and animal kingdoms. With the passage of time and numerous generations, the processes of favorable mutations and natural selection began to produce diversity. What was once a world limited to animals of one body type became a world with two. Then three. And so on, until the passage of eons and the process of evolution brought into existence every species known today.

single-cell organism → multi-cell organsim → fish → amphibian → reptile → bird / mammal

This is what evolution claims, but it is not what the early geological evidence shows. One or more members from each animal phyla (major grouping) suddenly and inexplicably appeared, fully developed, in the Cambrian period. They appeared as inexplicably as did the Mighty Speck nineteen billion years earlier.

Cambrian citizenry included a variety of organisms such as starfish, jellyfish, worms, sponges, and vertebrates. As these animals are of the same approximate age, it is illogical to postulate that one phylum gave rise to another. They were contemporaries. They lived side by side, not one after another. Where are the tens of thousands of fossils one expects to find between Precambrian one-cell and Cambrian creatures? They do not exist. Millions of fossils have been found to date—tens of thousands of transitional fossils anticipated, yet none found.

> *To the question why we do not find rich fossiliferous deposits belonging to these assumed earliest periods prior to the Cambrian system I can give no satisfactory answer.*[19]
> —CHARLES DARWIN, *The Origin of Species*

For good reason the Cambrian Explosion is an embarrassment to evolutionists. It alone is enough to deal a death-blow to the theory.

The Evolutionary Tree (Phylogeny)

Phylogeny is the arrangement of organisms into an evolutionary history. It assumes evolution is true, and fits (oftentimes with the help of a little force) the data into a Darwinian mold. A diagram in the shape of a single-trunk tree with many branches is used to depict the progression of evolution, with less complex species near the bottom of the tree trunk and more complex species at the tips of the branches. Man takes the summit position, with our first cousin, the monkey, swinging from a nearby limb.

Evidence suggests the evolutionary tree is a bit out of whack. Rather than a single-trunk tree with many branches, the Precambrian-Cambrian fossil record indicates something altogether different. Because all animal phyla appeared at the same time with no indication they derived from a common origin, it is misleading to use a single tree. What is needed is one tree for each phylum, as each one arrived on earth fully formed and without predecessor. The data call for an evolutionary forest of slightly branched trees, not one highly branched tree.

The Cambrian Explosion on Campus

Curious to know what university texts are saying about the Cambrian Explosion, I again turned to the 1300-page writ, *Biology*, which provided a brief yet candid description (emphasis mine):

> Rocks rich in fossils represent the oldest subdivision of Paleozoic era, the Cambrian period. From about 565 mya to 525 mya, *evolution was in such high gear*, with the sudden appearance

of many new animal groups, that this period has been nicknamed the Cambrian explosion. Fossils of *all contemporary animal phyla are present*, along with many bizarre, extinct phyla, in marine sediments. The sea floor was covered with sponges, corals, sea lilies, sea stars, clam-like bivalves, primitive squidlike cephalopods, lamp shells (brachiopods), tribolites. . . . In addition, small vertebrates—cartilaginous fishes, first reported in 1999—became established in the marine environment. Scientists have *not* determined the factor or factors responsible for the Cambrian Explosion, which has been unmatched in the evolutionary history of life.[20]

Evolution was in "high gear"? More like Warp 10! A basic tenet of evolution is that less complex species evolve into more complex species through incremental improvements over thousands of generations and millions of years, yet every phylum appears simultaneously in the Cambrian period, with not even one of those phyla leaving a trail back to, or anywhere near, the supposed original organism.

> *[N]evertheless, the difficulty of assigning any good reason for the absence of vast piles of strata rich in fossils beneath the Cambrian system is very great.[21]*
> —CHARLES DARWIN, *The Origin of Species*

Here is a skeptic's look at the types of things high-gear evolution zooms past. An animal with a skeleton (shell) on the outside evolved into one with a skeleton (backbone) on the inside. The monster halfway in between vertebrate and invertebrate would no longer have an outer shell for protection and support (since the shell has moved 'inside' the body), but it would not yet have a backbone for support (still waiting on the migrating shell to show up, turn to bone, and grow nerves). No supporting intermediate fossils have been found to support the theory despite the fact that shells and bones are excellent materials from which to form fossils.

Or perhaps fish evolved from an animal that was soft through and through, such as a jellyfish. Again the fossil record falls short of providing support for such a transition. The idea that a creature with soft inner parts formed bones that somehow managed to connect to the brain and serve as the conduit for a complicated neural network stretches credibility beyond the breaking point. Fish appear in the fossil record suddenly and fully formed.

> *If numerous species, belonging to the same genera or families, have really started into life at once, the fact would be fatal to the theory of evolution through natural selection.*[22]
> —CHARLES DARWIN, *The Origin of Species*

The biology textbook candidly admits that "Scientists have not determined the factor or factors responsible for the Cambrian Explosion." This appears to be a real stumper, given Charles Darwin said much the same thing 150 years ago.

~ 16 ~

IRREDUCIBLE COMPLEXITY

The Case Against Darwinian Fuzzy Logic

In his book *Darwin's Black Box,* biochemist Michael Behe presents the concept of *irreducible complexity.* He makes the case that complex machines and biological systems are all-or-nothing propositions: they must have all parts in their proper place before the thing as a whole will work.

Behe draws the analogy between complex biological systems and a mousetrap. If any of the pieces are missing or misaligned—the catch that holds the cheese, or the spring that throws the hammer, or the holding bar that holds the hammer back, or the platform that supports and precisely aligns all the parts—the mousetrap will not work. The issue is not that it will work poorly; rather, that it will not work at all.

Like the mousetrap, biological systems are all-or-nothing propositions. Consider, for example, the eye. Possession of half the parts of an eyeball does not give poor vision. It gives no vision. All parts must be present, operable, and correctly interconnected or there is total blindness. The complexity of the eye cannot be reduced and still work: it is irreducibly complex.

How did the non-functioning eye manage to stay in the head of species for the numerous generations it took to form a working eye? What advantage allowed the blind eye to escape the death axe of natural selection for so long? If the blind eye (or any of its parts) served a purpose in developing species, what was it? Serious answers are lacking. One can only imagine the deleterious affects of a non-functioning hole in the head.

> *To suppose that the eye with all its inimitable contrivances for adjusting the focus to different distances, for admitting different amounts of light, and for the correction of spherical and chromatic aberration, could have been formed by natural selection, seems, I freely confess, absurd in the highest degree.*[23]
> —CHARLES DARWIN, *The Origin of Species*

According to the theory of evolution, a deformity (variation through mutation) doesn't stay in the gene pool unless it offers an *immediate* survival advantage. Evolution doesn't keep a non-functioning eye because it *might* be useful in the future. Evolution lacks foresight. It operates under a "use it or lose it" rule.

Irreducible complexity is the eight-hundred-pound gorilla in the corner. Evolutionists continue to invent fanciful scenarios of common descent, ignoring irreducibly complex systems that separate the closest of phylogenic relatives. Horse-like animals give rise to giraffe. Sea animals that swim give rise to land animals that walk (fish to amphibian evolution), and land animals that walk give rise to sea animals that swim (land mammal to whale evolution). No scenario is too far fetched; too bizarre.

Imaginative scenarios such as these are what Michael Behe might describe as "fuzzy word-pictures typical of evolutionary biology."[24] He jests to the effect that a fuzzy word-picture could be constructed to describe how his daughter's toy fish could be changed "step by Darwinian step" into a Mississippi steamboat.[25]

Irreducible complexity is problematic to Darwinism. Even the most ardent supporter of Darwin's theory recognizes that a large number of biological systems are far too complex to have formed in one fell swoop, yet there is no viable mechanism to explain how such systems could have evolved gradually over time. Without biological function during each and every step of eye development, for example, there is no immediate survival advantage, and the Darwinian chain of progress through successive generations is broken.

Actually, irreducible complexity is much more than problematic to Darwinism. It is fatal.

> *If it could be demonstrated that any complex organ existed, which could not possibly have been formed by numerous, successive, slight modifications, my theory would absolutely break down.*[26]
> —CHARLES DARWIN, *The Origin of Species*

~ 17 ~

MACROEVOLUTION AND MICROEVOLUTION

We Must Come to Terms

If asked, "Do you believe in evolution?" I would be forced to reply something like, "It depends on what you mean by evolution." I would have to do so because the honest answer could be either "yes" or "no."

Semantics relating to the term *evolution* is more complicated than most realize. The term means different things to different people or in different situations. A couple of factors complicate the semantics. (1) There are two types of evolution. (2) The term evolution has both generic and specific meanings. When discussing evolution, it is *very* important to know which type of evolution is being discussed.

Two Types of Evolution

When John and Jane Doe think of evolution, they typically have the idea of one species changing to another. Amoeba-to-man. Fish-to-philosopher. Scientists refer to this as *macroevolution*. It refers to the process of evolving more complex organisms from less complex organisms, as evidenced by the production of increasingly complex body structures, new organs, advanced molecular machines within the cells, and novel body plans (body designs).

Macroevolution is to a biologist what a major retrofit is to a mechanic. Using a bicycle metaphor, our mechanic upgrades the frame, adds an engine, re-routes the chain, connects a fuel tank to the engine with rubber tubing, and adds a throttle to the handlebar. The original body plan has given rise to one that is genuinely different: a bicycle has given rise to a moped. Transportation has evolved.

In contrast, there is a type of evolution called *microevolution*. It deals in tweaks, rather than retrofits. A white bicycle is turned into a black bicycle; a bicycle with narrow handlebars is turned into a bicycle with wide handlebars. Minor changes (tweaks) are made, but the basic body plan does not change. We start with a bicycle and end with a bicycle.

In the real world, microevolution reveals itself in superficial changes to a species, such as alterations in the shape, size, or color of body parts, without changes in the basic body plan of the organism. There are big dogs and little dogs, cute dogs and ugly dogs, yet all are dogs. There is everything from very short jockeys to very tall basketball players, from narrow-framed models to large-framed sumo wrestlers, yet all are people. Or one might find variations among a given species of moths, some being light-colored whereas others are dark-colored, yet all are moths. Some finches have wide beaks and others narrow, but all are finches. Horses vary from 75-pound pygmy horses to 2,000-pound Clydesdales. There is enough latitude within a species' DNA pool to allow for great changes in shape, size, and color.

The following chart helps summarize the differences between microevolution and macroevolution.

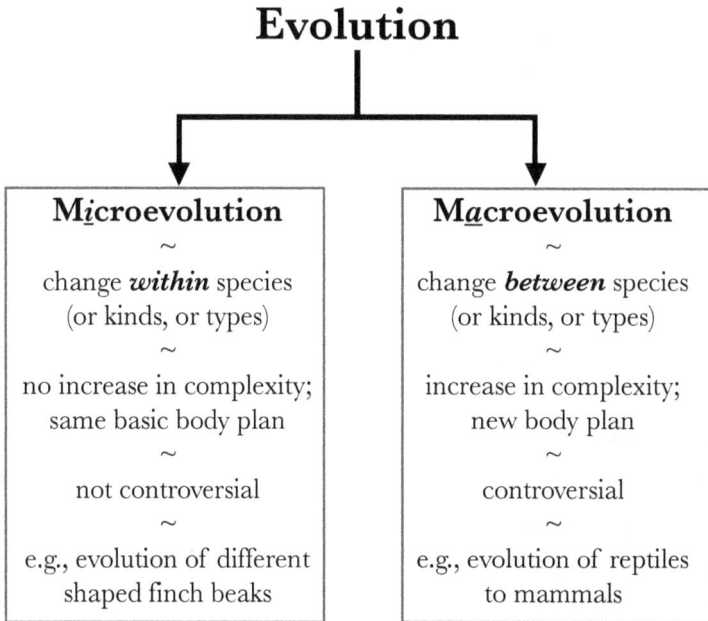

Evolution

Microevolution	Macroevolution
~	~
change *within* species (or kinds, or types)	change *between* species (or kinds, or types)
~	~
no increase in complexity; same basic body plan	increase in complexity; new body plan
~	~
not controversial	controversial
~	~
e.g., evolution of different shaped finch beaks	e.g., evolution of reptiles to mammals

Macroevolution is controversial, whereas microevolution is not. There is common ground among all scientists that species can and do evolve in the microevolution sense. In fact, the principles behind microevolution are put into practice every day: breeders use *variation and selection* (albeit artificial selection rather than natural selection) to produce leaner cattle, faster horses, insect-resistant crops, and prettier flowers.

The problem lies with macroevolution. Modifying or tweaking an existing species (microevolution) is easy; making a new species (macroevolution) is an altogether different situation.

The Species Problem

Before proceeding with the controversy over macroevolution, it is necessary to take a close look at the term *species*. Given evolution's main goal is to explain the appearance of new species (a process referred to as *speciation*), it is important to know what a species is. But what exactly is it? At what point is one species recognized to have evolved into another? Are species best defined by body structure (morphology), genetics (molecular biology), or some other criteria?

The concept of *species* falls within the field of taxonomy, the science of classifying organisms. Because there are many ways to slice and dice classification data, biologists employ a number of taxonomic systems. The most familiar system, developed by Carolus Linnaeus, the father of modern taxonomy, employees seven major taxonomic levels. *Kingdom* is the highest (most general) level; *species* is the lowest (most specific) level. The following table provides examples of taxonomic levels for several types of familiar organisms: humans, cats, and dogs.

When one refers to a species, both genus and species names are typically provided in italics, with the first letter of the genus capitalized. For example, your dog Spot belongs to the species *Canis familaris*. Your pet lion Leo belongs to species *Pantherina leo*. Wile E. Coyote belongs to genus *Canis lantrans* (not *Eatius birdius*, as he once claimed). Because Linnaeus lived in the 1700s before Darwin's time, there was no pressure to force taxonomic data into a phylogenic (common ancestor) tree. Linneaus was free to reflect nature as it reveals itself: in distinct groupings or types. (You will see in a later

LEVEL	HUMAN	DOG TYPES	CAT TYPES
KINGDOM	Amimalia	Amimalia	Amimalia
PHYLUM	Chordata	Chordata	Chordata
CLASS	Mammalia	Mammalia	Mammalia
ORDER	Primates	Carnivora	Carnivora
FAMILY	Hominidae	Canidae	Felidae
GENUS	Homo	Canis	Felis (house cat) Pantherinae (lion) Acinonyx (cheetah)
SPECIES	sapiens	familaris (dog) lupus (wolf) latrans (coyote)	catus (house cat) leo (lion) jubatus (cheetah)

A mnemonic device to help you remember the seven major levels:
"Keep Plates Clean, Or Family Gets Sick"

chapter examples from molecular biology that support Linneaus' approach to taxonomy.)

The term *species* typically refers to a population "whose members are capable of interbreeding . . . and do not interbreed with members of other species."[27] I say typically because there are a number of definitions for species, none of which fit all situations. The inability to unambiguously define what a species is, and what its limits are, is referred to as the *Species Problem*. Depending on the situation, an evolutionist may "view" a species in any number of ways. There's the topological view, the morphological view, the phylogenic view, the cladistic view, the taxonomic view, the ecological view, and of course, the evolutionary view. An internet search for the term "species problem" provides a bountiful yet dizzying array of hits.

Even within the confines of the "different species don't interbreed" definition exceptions abound. For example, dogs (*Canis familiaris*), wolves (*Canis lupus* and *Canis rufus*), and coyotes (*Canis latrans*) are able to interbreed but nevertheless classified as different species. Of course, for extinct organisms, it is impossible to determine breeding limitations with certainty, so species assignments are necessarily based on opinion rather than objective criteria.

The problem is as old as Darwin, who wrote, "No one definition has satisfied all naturalists; yet every naturalist knows vaguely what he means when he speaks of a species."[28] To paraphrase, "I can't really define it, but I know it when I see it."

You probably see the big problem: the definition of species sets the standard for establishing what is (and is not) evidence for macroevolution, and yet there is no universal agreement as to the definition of *species*! For example, evolution may produce a change in organisms whose basic body plan remains unchanged (tenet of microevolution) but nevertheless cannot interbreed (tenet of macroevolution). Evolutionists can claim a new species has formed despite the fact that Darwinian evolution—as demonstrated by increasingly complex body structures, new organs, advanced molecular machines within the cells, or altered body plans—has not occurred.

Where does one draw the line between species? The answer seems to be, wherever the most influential taxonomist says to draw it.

In an attempt to avoid semantic pitfalls, I will occasionally use the term "body plan" or "kind" instead of "species." To borrow a riff: no one definition will satisfy all readers; yet every reader knows vaguely what I mean when I speak of a body plan or kind.

> *What's in a name? that which we call a rose*
> *By any other name would smell as sweet....*
> —WILLIAM SHAKESPEARE, *Romeo and Juliet*

Regardless of the term you use—species, kind, body plan—the key is to distinguish between macroevolution (an evolutionary "retrofit") and microevolution (an evolutionary "tweak"). Don't get too caught up in the name.

The Great Controversy: Macroevolution

Darwinists insist microevolution is just macroevolution in increments too small to be "macro." They believe microevolution is merely macroevolution baby steps. An evolutionist's chart of the relationship between microevolution and macroevolution looks similar to the one above, except for one critical difference: the flow of the arrows.

Evolution
(Darwinian)

Microevolution	Macroevolution
~	~
change *within* species (or kinds, or types)	change **between** species (or kinds, or types)
~	~
no increase in complexity; same basic body plan	increase in complexity; new body plan
~	~
not controversial	controversial
~	~
e.g., evolution of different shaped finch beaks	e.g., evolution of reptiles to mammals

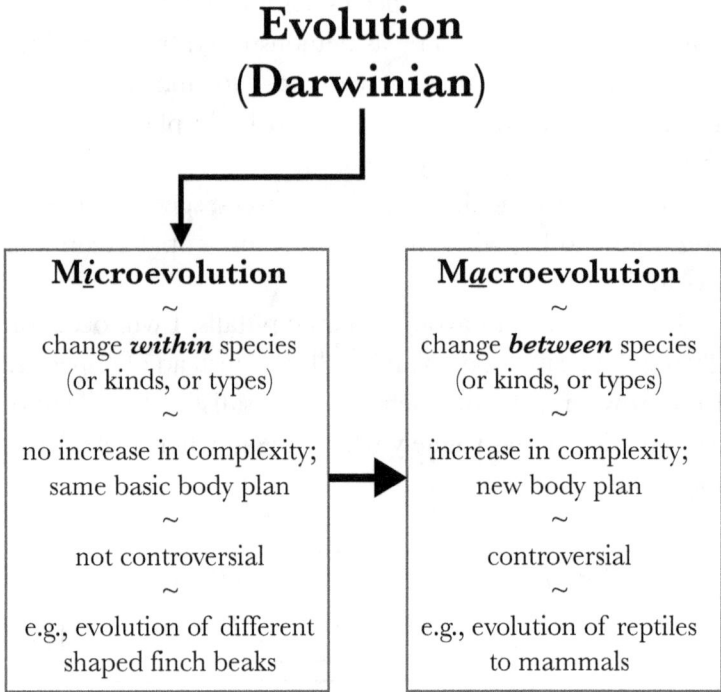

Whether microevolution and macroevolution constitute one smooth process (as evolutionists believe) or two independent processes that cannot be bridged (as anti-evolutionists believe) is an issue of utmost importance. Evolutionists cannot allow to stand any hint that microevolution and macroevolution are independent. They will fight to the death on this point. The reason: there is no known mechanism to explain macroevolution—no mechanism to explain the large-scale biological changes needed to evolve one species from another, or to overcome the obstacle of irreducible complexity. The only game in town (aside from Intelligent Design) is to claim macroevolution the cumulative result of microevolution.

But this game is a loser. It flies in the face of real-world data, relying instead on an illusionary interdependence between microevolution and macroevolution. As far back as Darwin's time, breeding experiments showed that species could be varied significantly, *but only within certain limits.* Genuinely new types of organisms were never produced. The bridge between micro- and macroevolution is nonexistent. Evidence since Darwin sings the same song: microevolution can and does vary *existing* biological structures, but it cannot create new ones. And without the formation of new and novel structures there can be no speciation (macroevolution).

Which Way is Up?
Horizontal Versus Vertical Evolution

Evolution is "a process of change in a certain direction."[29] The big question is, "What direction?"

This question has lead some to use the more descript terms *vertical evolution* and *horizontal evolution* in place of macroevolution and microevolution, respectively. They make explicit the *direction* of change. Vertical evolution refers to the type of evolution in which species are viewed as advancing up the evolutionary ladder, from less complex to more complex species. Horizontal evolution refers to the type of evolution in which species change is not accompanied by an increase in complexity. Species are changed, but they are not advanced. Structures, organs, and molecular machines are altered, but only so far—new ones are not added. Macroevolution moves species in the up direction (vertical evolution); microevolution moves them sideways (horizontal evolution).

If you want to spot microevolution (evolutionary tweaks) among the scientific smog, use the "same-but-different" rule.

If the beginning and ending organisms are the *same* kind (same species, same body plan, same structures, or same level of complexity) *but different* in superficial characteristics (different sizes, shapes, or color), then microevolution has probably occurred.

Actually, We Are All Evolutionists!

Confusion reigns when the term *evolution* is used to mean macroevolution one moment, microevolution the next, and both types the next. The word *evolution* usually implies macroevolution, but it is important to recognize it can also refer to microevolution.

Now, let me return to the question posed at the beginning of this chapter.

"Do you believe in evolution?"

"It depends on what you mean by evolution."

"I am referring to the evolution of different characteristics without a basic change in body plan or complexity."

"Yes, I believe in evolution."

However, the questioner could have led me to a different response.

"Do you believe in evolution?"

"It depends on what you mean by evolution."

"I am referring to one species evolving into a new and more complex species, or one with a noticeably different body plan."

"No, I do not believe in evolution."

Scientists who subscribe to macroevolution are referred to as *Darwinists*, to differentiate them from scientists who believe in microevolution but reject macroevolution. Evolution in the horizontal direction is fact; evolution in the vertical direction (Darwinism) is not.

Every scientist accepts horizontal evolution as true, including creation scientists. After all, creationists require that from a pair of animals come every shape and size of that kind of animal in the world today. For example, it is supposed that both gray and black moths descended from a lone pair of ancestors with enough latitude in their DNA to allow for great variety within the species. Creationists and evolutionists may disagree on the merits of macroevolution, but they find common ground in the concept of microevolution. We are all (micro)evolutionists!

While it is true that all scientists are evolutionists, it is not true that all scientists are Darwinists.

~ 18 ~

EVOLUTION IN ACTION

Bait and Switch

We are all familiar with the bait-and-switch sales tactic. It begins with the bait, an advertised product that appears "too good to be true" for the price. But amidst the wheeling and dealing, the salesman stealthily switches the bait with an inferior product. The customer proudly takes the "too good to be true" purchase home, only to realize the product bought is not the same as the product advertised.

Examples of microevolution are often cited as evidence of Darwinian macroevolution, sometimes under the banner "evolution in action." Unfortunately, non-scientists will often fail to recognize that an example of microevolution (horizontal evolution) is not an example of macroevolution (vertical evolution). The bait, a promised example of macroevolution, has been switched with an example of microevolution.

The Peppered Moth Story

There may be no better illustration of this than the tired and overworked story of the Peppered Moth.

In pre-industrial England, most of the peppered moth (*Biston betularia*) population was gray in color, which afforded effective camouflage when loitering about the gray tree trunks in that fair country. Black moths were few, as hungry predators could easily spot the black moths against the lighter background of the gray trees.

However, the arrival of the industrial revolution changed all that. It brought pollution of the black sooty sort, which collected on the gray tree trunks, turning them black. With the tree trunks now black, hungry predators began to preferentially spot, and subsequently eat, the gray moths. Black moths thus became dominant in the English moth population.

And there you have it: "evolution at work." Where there were once gray moths, now there are black. Apparently chance mutations produced black moths, and natural selection sustained them. Darwin must have been right. Right?

The peppered moth story has nothing to do with the type of evolution that produces new species. Darwin clearly made the case for macroevolution, whereas the moth change represents, at best, microevolution. The peppered moth story starts with moths, and it ends with moths. Neither complexity nor body plan was altered. No new organs or biological structures were produced. The mothy change was evolution in the microevolution sense, but not in the macroevolution sense. No mutations, a key component of Darwinian evolution, were required. England had, in the end, the exact same species with which it started: *Biston betularia*.

Actually, the case could be made that there was no evolution at all. Before the industrial revolution, both black and gray moths were present (but more gray ones); after the industrial revolution, both black and gray moths (but more black ones). Both types of moths predated the industrial revolution. What we really see is merely a shift in population abundance.

Peppered with Scandal

The peppered moth story turns from bad to worse to deceitful. The textbook photos of peppered moths resting on tree trunks: staged. Dead moths were glued or pinned to the trees, or docile ones were strategically placed by hand.

Remarkably, peppered moth apologists defend the arguably fraudulent practice. From the staff of The National Center for Science Education (NCSE), " . . . the important issue here is not how the photos were made, but rather their intent. . . . how the photos were produced does not change the actual data."[30] Why let a little fraudulence get in the way of a good story!

Try that logic in a court of law and see how well it stands up. Imagine the defense of a well-intentioned policeman on trial for using a throw-down weapon to frame a violent criminal. "The important issue, ladies and gentlemen of the jury, is not that my client used a throw-down pistol. What's important here was his intent: to take a vicious felon off the street." My guess is that someone is going to jail—and it's not the vicious felon.

Unfortunately, this is not an isolated event. For example, there is also great controversy surrounding Ernst Haeckel, a contemporary of Charles Darwin. Haeckel made misleading drawings of various embryos in support of Darwin's new theory. He drew them to look similar where, in reality, they are not. Modern photos of the same types of embryos

Haeckel drew demonstrate the audacity of the Victorian-age drawings. (For a real eye-opener, see the companion DVD to the book *Icons of Evolution.*) Nevertheless, the drawings are so compelling that biology textbooks have been using them for decades—as late as 2003!—to illustrate Darwinian evolution. NCSE doesn't seem particularly bothered at that either:

> The charge that Ernst Haeckel intentionally "faked" his drawings is irrelevant. Regardless of his intent, the drawings that Haeckel made are incorrect, especially in what he labeled as the "first stage." But it really does not matter what Haeckel thought or whether his drawings are accurate: modern comparative embryology does not stand or fall on the accuracy of Haeckel. . . .[31]

In addressing "the charge" of fakery, NCSE does concede the drawings are "incorrect." Some might argue that Haeckel's drawings were *incorrect* in the same sense Bernie Madoff's accounting practices were *incorrect*. (In 2009, Madoff was sentenced to 150 years in prison for *incorrect* accounting practices that went in his favor to the tune of about $20 billion—that's billion with a "b.")

Fraudulent practices, or charges thereof, notwithstanding, there are problems with the design of, and resultant data from, the original peppered moth experiments of biologist Bernard Kettlewell. After laying out the facts in his book *Icons of Evolution,* Jonathan Well's remarks:

> Industrial melanism [color change] in peppered moths shows that the relative proportions of two pre-existing varieties can change dramatically. This change may have been due

to natural selection, as most biologists familiar with the story believe. But Kettlewell's evidence for natural selection is flawed, and the actual causes of the change remain hypothetical. As a scientific demonstration of natural selection— as "Darwin's missing evidence"—industrial melanism in peppered moths is no better than alchemy.[32]

Darwin's Finches

One of the most famous examples of "evolution in action" is Darwin's finches. During his five-year trip aboard the *H.M.S. Beagle*, naturalist Charles Darwin visited the Galapagos Islands, located off the coast of South America. He was impressed that Galapagos and South America finches had many similarities yet important differences, including beak size and shape. Darwin claimed the finches evolved from a common ancestor; differences among finches were attributed to adaptations necessary for survival among the various ecological habitats of the islands. Big-beak finches were best at cracking seeds, and slender-beak finches could ferret insects.

There is no question that finches undergo microevolution when the environment is altered and survival is at stake. In his book *Darwin on Trial*, Philip Johnson cites the case in which a 1977 drought in the Galapagos Islands caused a noticeable shift in the average size of the finches. The drought brought small seeds into short supply. A disproportionate number of smaller birds, unable to eat larger seeds, died out in favor of larger finches.[33]

Darwin's Finches provide an example of *horizontal* evolution. Natural selection preserves the fittest finches (in this case, those with larger beaks) at the expense of less fit finches. But nevertheless, finches one and all.

However, when the drought went away, so did the larger beaks. Finch beaks returned to their pre-drought size with the return of the rains. The evolutionary change wasn't permanent, yet evolution is supposedly an irreversible process.

Darwin's finches may now be one of the darling examples of evolution, but such was not always the case. In *The Origin of Species*, Charles Darwin failed to make the connection between finch beaks and evolution. They didn't even make honorable mention.

Galapagos finches, English peppered moths, and animals worldwide vary from generation to generation, and natural selection preserves the fittest; however, the resulting variation is always limited to the horizontal direction.

> None of the "proofs" provides any persuasive reason for believing that natural selection can produce new species, new organs, or other major changes, or even minor changes that are permanent.[34]

If the moth and finch stories are mere examples of microevolution, where, then, are the examples of true macroevolution? Nowhere to be found.

> The most fundamental problem of evolution, the origin of species, remains unsolved. . . . What Darwin claimed is true for all species [macroevolution] has not been demonstrated for even one species. Trying to finesse the problem by calling minor changes within existing species "microevolution" is like trying to over-

come obstacles to space travel by calling a tod-
dler's first footsteps "microastronautics."[35]

~ 19 ~

MULTI-CELL ANIMALS TO FISH

More Holes than a Donut Factory

By the time the Ordovician period arrived, continents had drifted apart and large bodies of water were lavish with the Cambrian creatures. Somewhere in time, one of those aquatic dwellers was born with a bump. The bump was passed along to descendents, eventually mutating into a lump. The passing of time and generations would see the lump evolve into a flattened and moveable superlump and, eventually, the superlump into a fin. The bump-to-lump-to-fin scenario occurred about the fish's body at multiple points (fish tend to have more than one fin) simultaneously (equal-size fins) and symmetrically (fins complement one another). Were evolution to have randomly placed fins willy-nilly about the fish's body, the creature might be limited to, say, swimming in circles in perpetuity—hardly an evolutionary advantage.

Of course, I have no proof for my bump-to-lump-to-fin mechanism—I'm making it up as I go—but herein lies the beauty of the theory of evolution. It accommodates just about any Toy-Boat-to-Mississippi-Steamship scenario imaginable.

One might challenge my theory with the question, "What about the lack of transition fossils with bumps, lumps, and superlumps?" I would reply there aren't any because evolution was in high gear at the time. Again, quoting from *Biology* (emphasis mine):

> In the Ordovician period, much land was covered by shallow seas, in which there was *another burst of evolutionary diversification*, although not as dramatic as the Cambrian explosion.[36]

Reminiscent of the Cambrian explosion, fish suddenly appeared on the earth in a fully formed state.

First, there are gaps between one-cell and simple organisms; now, gaps between simple organisms and fish. If this trend continues, the fossil record is going to have more holes than a donut factory.

TAUTOLOGIES

The Cat Chasing Its Tail

Someone makes the statement, "Too much of a good thing is bad for you."

Deciding to get to the bottom of the situation, you ask, "If it's a good thing, why is it bad?"

"Because it's too much."

"How do you know it's too much?"

"Because it's bad for you."

"On what basis do you say it is bad?"

"On the basis that it's too much."

And so the dialog can go in circles, *ad infinitum*. Thoughts are confounded. It's bad for you because it's too much, and it's too much because it's bad for you. One thing ("too much") points to the other ("is bad"), and vice versa. It is difficult to see where the logic is flawed, for something that is too much is in fact bad.

This logical mess is called a *tautology*, better known as circular reasoning. The same idea is being expressed twice, but in different words. Here are a few Yogi Berra tautologies, just for fun:

> "It's *deja vu* all over again."
> "We made too many wrong mistakes."
> "If you don't get any better, you'll never improve."
> "You wouldn't have won if we'd beaten you."
> "You can observe a lot by just watching."
> "It ain't over till it's over."
> "A nickel ain't worth a dime anymore."
> "I knew my record would stand until it was broken."

Being redundancies, tautologies are necessarily true. Tautologies such as "Yogisms" are amusing to consider, but tautologies pose a problem when scientists employ circular reasoning in an attempt to prove a serious point.

Fossil-Rock Tautology

Evolution teaches that less complex animals evolved earlier in geologic time; hence, their fossil remains are expected to lie deep in the geological column, within the oldest sediments. On the other hand, sediments that contain the fossils of more complex creatures are considered relatively new. Younger fossils are expected to lie higher in the geological column; in shallow graves, if you will.

Geologists sometimes use so-called *index fossils* to determine the age of some rocks. These fossils are "known" to have existed in a certain geological era. If a rock has a 500-million-year-old index fossil engraved in it, then it must obviously be at least 500 million years old. This sounds logical, but consider the following conversations.

"Hey, look at this *rock* formation! How old is it?"

"500 million years old."

"How do you know?"

"Because it has a 500-million-year-old fossil in it."

The next day, same location, same rock, but two different people talking:

"Hey, look at this *fossil!* How old is it?"

"500 million years."

"How do you know?"

"Because it is contained within a 500-million-year-old rock formation."

The fossil defines the age of the rock, and the rock defines the age of the fossil. Welcome to the world of circular reasoning.

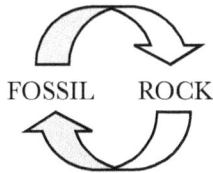

FOSSIL   ROCK

This example relies on the assumption that the simplest organisms are the oldest. It assumes evolution is true.

Of course, not all rocks are dated using index fossils. They may be dated, for example, using radioactive dating (which has its own set of assumptions). But where index fossils are concerned, it's no wonder much of the geological data support the theory of evolution. By definition, it has to—it's tautologous.

Natural Selection as a Tautology

Some have argued that natural selection—a.k.a. survival of the fittest—is a tautology.

In evolution parlance, the "fittest" is that which leaves viable offspring. A smart wimp may be more fit for survival that a thickheaded brute of great strength. Dinosaurs and Neanderthals have given way to computer hackers and chemists with pocket protectors.

"Who will survive?"

"Those who are most fit."

"Who are the most fit?"

"It's those that survive."

"But why do they survive?"

"Because they are the most fit."

The fittest are those that survive, and those that survive are the fittest.

The
FITTEST

The
SURVIVORS

This statement is necessarily true by definition. Because Darwinism requires that the more fit survive (and the less fit go extinct), there is no way for the statement to be false. As presented, it is a tautologous statement that merely tells us what we already know: survivors survive. Thanks for the news flash.

Others will argue, with limited merit, that the phrase should not be taken too literally, that there is more to "fitness" than mere survival. "Survival of the fittest is merely a catch phrase," they might protest, "it's not the whole story."

Tautology or not, the term *natural selection* is nevertheless convenient. In two words, it brings to mind a process by which nature (as opposed to an entity with intelligence) selects (preserves or renders extinct) something that already exists. Scientists as a whole do not have a problem with the concept that nature picks winners and losers. However, they disagree mightily when it comes to the issue of how the organism on which natural selection acts arose about in the first place. Aye, there's the rub.

PART FOUR
Fish to Philosopher

~ 21 ~

FISH TO AMPHIBIANS

The Oldest Fish Tale in the World

Fish evolved about ninety-eight percent of the way through history, evidence that life forms had climbed the evolutionary ladder rather nicely since the Cambrian Explosion. An evolutionary ladder, roughly in tune with the geological time scale, is shown in the following table.

Theory has it that the fish fin evolved into something like a short appendage with a fin on the end of it. A fish so endowed, realizing that the appendage possessed forward and backward movement and the fin-end could be used to support the body, decided to take a walk on the wild side, to go where no fish had gone before: dry land. We hairy ape-like mammals take walking for granted, but back then, walking on dry land was a big deal. It was the fish equivalent to modern-day man walking on water. Walking on dry land was a profound advancement that changed the fate of the world. It opened vast, new resources to the animal kingdom.

Rung	*Event*	*Comment*	*MYA*
10	Monkey to Man	We made it!	1
9	Reptile to Monkey	Hairy first cousins	40
8	Reptile to Bird	Taking flight	200
7	Amphibian to Reptile	Scaly things	350
6	Fish to Amphibian	Land Ho!	400
5	Multi-cell to Fish	Ocean dwellers	500
4	One-cell to Multi-cell	Nature on a growth spurt	540
3	Abiogenesis	Life from the nonliving	2,600
2	Big Bang	Everything from nothing	20,000
1	Nothing		

MYA = millions of years ago

Once the walking fad caught on, it was a simple matter of the appendage evolving into a leg, and the fin into a foot. The new land-walkers were dubbed amphibians.

A Historic Catch

But what of the creatures halfway between fish and amphibians? One of the amphibian ancestors was the coelacanth (see-la-kanth). Fossils show it to have had leggy-fin type appendages, obviously leg-foot precursors.

As the internals of amphibians are different from fish, it is supposed that the internal organs of the coelacanth were probably somewhere between those of fish and modern-day amphibians. No one will ever know for sure, as the coelacanth became extinct some 70 million years ago. Or did it?

In December of 1938, a trawler fishing off the coast of South Africa caught a live coelacanth. Alive and well! (At least it was before the fisherman caught it.)

What a great opportunity for evolutionists! The unfortunate coelacanth provided an opportunity to prove to the world the existence of a true intermediate form, a chance to prove that the predictions of Darwinists are more than fuzzy word-pictures typical of evolutionary biology.

The five-foot South African fish/amphibian creature was gutted with care and thoroughly inspected. And what was discovered? The leggy-fin appendage turned out to be all fin, and the internal organs turned out to be all fish. It was a fish, a whole fish, and nothing but a fish.

With no way to disprove them, highly speculative just-so scenarios that inundate evolution are given the benefit of the doubt. But in this case, we find an evolutionary 'just-so' story that is just not so.

Mosaics

O.K., the coelacanth turned out to be a dud. But what about the lungfish? It has gills and fins like a fish, but a lung and heart like an amphibian. Isn't this a missing link? It seems half one thing and half another.

Rather than a missing link, the lungfish is a *mosaic*. To get a sense of what is meant by mosaic, consider the platypus. It has fur like a mammal, lays eggs and has webbed feet like a bird, carries poisonous venom like a reptile, lives like an amphibian, and finds food using electrical impulses like some fish. The platypus appears to have been made *a la carte* from fully functional body parts characteristic of many types of animals. It is not surprising that upon first seeing a platypus from Australia, Britons mistook the strange creature as a fake. Maybe someone had sewn a duck's bill onto a mammal's body. Even when accepted as real, the mammal was confused with a bird or reptile because it lays eggs.

A platypus is neither a "living fossil" nor an intermediate species. It does not have body parts that are half one thing and half another. Its body covering is not halfway between fur and scales, or fur and feathers—it is all fur. All of its body parts are fully developed, functional, and identifiable.

Likewise, the lungfish does not have a breathing apparatus that is one-half gill and one-half lung. It has gills that are one-hundred percent gills, and lungs that are one-hundred percent lungs. The lungfish is merely a mosaic composed of fully functional, recognizable body parts.

Betting on the Come

Evolutionists claim intermediates lie between every major group, some intermediates more convincing than others.

One of the amphibian-to-reptile gap fillers is a fishlike amphibian, *Acanthostega gunnari*. Unlike the coelacanth, an *Acanthostega gunnari* specimen has not been caught.

But even if candidate missing links such as *Acanthostega gunnari* and the coelacanth (had its true identity not been discovered, and the previous fallacious analysis remained intact) are given the benefit of the doubt, their basic body plans are still vastly different from the most closely related amphibian. They bear the same cross as supposed missing links from across the phylogenic spectrum: numerous irreducibly complex systems separate them from their closest phylogenic ancestors and descendents.

Darwinists await the discovery of a new principle or law of biology to explain the chance development of irreversibly complex systems, and the chance appearance of life itself.

Meanwhile, discussions on the merits of existing missing links will border on meaningless as long as closest phylogenic neighbors are separated by chasms. For a group of hikers, a discussion as to whether the gap before them is as wide as the Grand Canyon or five football fields is academic—they are not walking to the other side. Their only option: take a different path.

(Intelligent Design awaits.)

~ 22 ~

BENEFICIAL MUTATIONS

A Fly in the Ointment

Beneficial mutations, which provide the changes on which natural selection acts, are the magic wand of evolution. From a supposed single-cell species came mutants that float, swim, slither, walk, run, hop, and fly; eat plants and eat animals; lay eggs and live birth; weigh less than a gram to more than five tons. The basic idea is that, every so often, a "good" mutation (good deformity, if you will) occurs that imparts survival advantages. The advantaged mutant survives at the expense of less advantaged competitors. If the mutation is extensive enough to have produced a new species, *speciation* is said to have occurred.

The issue is not whether mutations occur. They can and they do. We have all heard about man-made casualties such as the victims of the 1986 Chernobyl nuclear disaster, or monstrosities such as two-headed cows and sheep. Rather, the issue is whether *favorable* mutations or good deformities occur and are propagated in nature.

Maybe you have never seen nor heard of a good deformity, but Darwinists assure us history is replete with them, and that they exist in our very midst.

Mutant Fruit Flies

One hailed proof of Darwinian evolution is the formation of a fly with two pairs of wings (a fly normally has one pair). Irradiating countless fruit flies with X-rays actually produced an extra pair of wings. Darwin himself would likely have been proud of the confirmation.

It is unlikely, however, that the fly was too impressed. Its newly evolved wings did not come with the muscles necessary to use them for flight. To make matters worse, the wings developed at the expense of two wing-like structures that help stabilize flight. Writes Jonathan Wells of the poor creature:

> In aerodynamic terms, a triple-mutant four-winged fruit fly is like an airplane with an extra pair of full-sized wings dangling loosely from its fuselage. It may be able to get off the ground, but its flying ability is seriously impaired.[37]

The pitiful four-winged fly ranks with two-headed snakes and other oddities unfit to survive in nature. Isn't Darwinian evolution supposed to produce species that are *more* fit to survive in nature, not less?

Sickle-Cell Anemia and
Antibiotic Resistant Bacteria

The development of sickle-cell anemia in people and antibiotic resistant in bacteria somehow make the list of convincing examples of beneficial mutations. (Yes, the words *beneficial* and *sickle-cell anemia* are supposed to be in the same sentence.)

Sickle-cell's advantage manifests itself during outbreaks of malaria. Those who carry the hereditary trait for sickle-cell disease are slightly less likely to die from malaria than those without the trait. On the other hand, 80% of those who contract sickle cell will die by the time they are young adults, amounting to multiple thousands of casualties per year. Sickle-cell is clearly detrimental to the population as a whole. For some reason, I don't think this is the type of "beneficial" mutation Darwin had in mind.

Neither is antibiotic resistance in bacteria. To achieve macroevolution, mutations must change an organism's morphology—it's form and structure.[38] However, the mutations that promote antibiotic resistance do not alter the bacteria's morphology.

The fact that sickle-cell disease, antibiotic resistance, and debilitated fruit flies remain among the crown jewel examples of beneficial mutations provides an indication of the poor state of the evidence. To be sure, these examples will be joined by other supposed examples, most of which will fall within the same general description: trivial, esoteric, and void of changes in basic morphology.

What you are unlikely to see are mutations that demonstrate the type of morphological changes that drive speciation; the type of mutations that produce four-winged fruit flies—that actually fly.

~ 23 ~

NATURAL SELECTION

A Tall Tale

Were they to actually occur, favorable mutations would be preserved in the gene pool through natural selection. The more favorable the mutation, the more fit the mutant, the greater the chances nature will select it to survive and leave offspring. In Darwin's own words, "This preservation of favorable individual differences and variations, and the destruction of those which are injurious, I have called Natural Selection, or the Survival of the Fittest."[39]

Darwin Follows in Spencer's Footsteps

Surprising to many, Charles Darwin did not coin the phrase "survival of the fittest," philosopher and biologist Herbert Spencer did. Even more surprising, Spencer laid down the basic concepts of the theory of evolution years before Darwin published *The Origin of the Species*! From Spencer's essay entitled "The Developmental Hypothesis:"

> [T]en millions of varieties have arisen by successive modifications. . . . [S]upporters of The Development Hypothesis . . . can show that any existing species—animal or vegetable—when placed under conditions different from its previous ones, immediately begins to undergo certain changes fitting it for the new conditions. They can show that in successive generations these changes continue; until, ultimately, the new conditions become the natural ones.[40]

Spencer clearly expresses the main tenets of what would eventually be branded the Theory of Evolution.

The originality of Darwin's work is often overplayed, leaving student's with the false impression Darwin was the first to come up with the idea of evolution—as if he were cruising around on the *H.M.S. Beagle* and, while pensively reflecting on the day's observations, suddenly had an epiphany. On closer inspection, it appears Darwin neither conceived of, nor "discovered," evolution. Darwin did an excellent job of collating the ideas of others (including those of his grandfather, and Herbert Spencer, and others) and blending them with his own in a comprehensive written record.

Anyway, Darwin preferred the term "natural selection," which made clear he was referring to a natural process driven solely by chance, divorced from any intelligent agency.

And the Wheels on the Theory
Go Round and Round

Variation refers to the genetic-driven, mutative process by which one competitor becomes different from the other, and *natural selection* refers to the process in which nature picks a winner between different competitors. In simplified form:

$$\text{Evolution} = \text{Variation} + \text{Natural Selection}$$

These two distinct processes, variation and natural selection, repeat over and over and over, turning as a wheel which rolls through the expanse of time, churning out new species as it goes.

Natural Selection: Powerless to Create

Natural selection is sometimes described as a "creative force," but this is simply not the case. Natural selection only has the power to preserve, or render extinct, that which has already been created through reproductive variation. If there is to be a creative force in evolution, it is reproductive variation by mutation, not natural selection. The big question, then, is how did the thing on which natural selection acts come about?

As always, one should be aware of semantics. Natural selection is sometimes taken to mean selection *plus* variation. Under that usage, "natural selection" could be considered a creative force by virtue of its inclusion of reproductive variation.

How the Giraffe Got Its Long Neck

To illustrate how natural selection works, consider the oft-told story of how the giraffe got its long neck. Apparently, those wonderfully long-necked vegetarians used to have short necks.

Eons ago, so the story goes, a horse-like animal was born with a deformed neck. Its neck was too long—but that turned out to be a good thing. It could reach higher than its peers for vegetation, thus assuring itself of a tasty meal during the worst of times. Hence, the long-necked creatures survived while their shorter-necked brethren died out. The long-neck variant did exceptionally well until a creature with an even longer neck evolved and survived in its stead. The scenario of longer-neck generations evolving and shorter-neck generations dying out repeated itself until a creature twenty feet tall had evolved. The giraffe entered the winner's circle in the Darwinian game of natural selection.

Despite its simplicity, there are a few nagging problems with the giraffe story. If shorter-neck giraffes died out in favor of longer-neck giraffes, and if female giraffes have necks two to three feet shorter than male counterparts (they do), then why didn't female giraffes die out long ago? And why didn't all animals with eating habits similar to the giraffe go extinct along with shorter-neck giraffes?

The unbelievable decades-old eating habit story has been replaced with a tale based on "necking." No, no, it's not what you think. Necking refers to combat between male giraffes where they club each other with their necks and head in the attempt to establish dominance. Since giraffes with longer and stronger necks tend to win, they get the girl, and their progeny survive.

Regardless of which just-so story is in vogue, if giraffes evolved from shorter-neck variants according to the principles of natural selection acting on variations over numerous generations, then many of the anticipated hundreds of intermediate species should be available in the fossil record. In a detailed report, geneticist Wolf-Ekkehard Lönnig rebuts the assertions of several advocates of giraffe neck evolution, making special mention of the lack of evidence:

> Gaps exist between all the genera of the Giraffidae, and not a single one of the numerous postulated "speciation events" has been preserved (granted that they ever occurred).[41]

The story of the giraffe is infinitely more complicated than a mere stretching of legs and neck. Because the giraffe's head is so high above ground level, unusually high pressures are needed to push blood up to the brain. But like makeup, too much of a good thing can be bad. When the giraffe bends its head down to drink, high blood pressure could cause brain damage. And perchance the thirst-quenched giraffe survive the drink, the sudden lifting of the head could cause an insufficient amount of blood to reach the brain. To remedy these problems, the giraffe has a complex network of check-valves and elastic vessels that accommodate a heart that can weigh over twenty pounds—and that's a big heart. The average human heart weighs less than a soccer ball, the average giraffe heart more than a bowling ball.

The irreducibly complex cardiovascular system of the giraffe and the lack of transitional species in the fossil record argue against a Darwinian explanation. Evolutionists are anxious to keep the giraffe story going, but Behe challenges them to explain, rather than merely claim, supposed transitions between species. He asks two questions of beetles, but they apply equally well to all animals:

> But the burden to Darwinians is to answer two
> questions: First, what exactly are the stages of
> beetle [or giraffe, or any other animal] evolu-
> tion, in all their complex glory? Second, given
> these stages, how does Darwinism get us from
> one to the next?[42]

The problem, then, is not with natural selection itself.
Natural selection works. In a dog-eat-dog world, pit bulls will
survive and Chihuahuas will not. Bigger giraffes will beat up
smaller giraffes.

*Two campers spot a bear running towards them.
In haste, one camper begins putting on his running
shoes.*

*"Why are you wasting time doing that?" asks the
other camper, "You can't outrun a bear."*

*"I don't have to," came the reply, "I only have to
outrun you."*

The issue in dispute centers on *how* the adaptations on
which natural selection acts arise in the first place. What
mechanism accounts for the extreme changes required to
turn cardiovascular system of a horse-like quadruped into
the cardiovascular system of a giraffe? Evolutionists have yet
to offer a biological mechanism that accounts for the development of even one irreducibly complex system, yet every
plant and animal is essentially an assemblage of irreducibly
complex systems.

It boils down to this: nature knows how to preserve a favorable adaptation, she just doesn't know how to make one.

~ 24 ~

AMPHIBIANS TO REPTILES

A Cat Burglar's Path

Scenarios as to how amphibians came to be reptiles are a bit dodgy, at best. In addition to the glaring lack of evidence in the fossil record, there are huge differences between reptiles and amphibians that could use some splainin', Lucy.

One of the most notable differences lies in their reproductive systems. Reptiles use hard-shelled eggs suited for dry environments (amniotic eggs), whereas amphibians use jelly-like eggs suited for water. But there's much more to this story than slapping a hard coating on a soft amphibian egg and calling it a reptilian egg. Michael Denton lists a number of innovations necessary to accomplish the task (numbers added):

> Altogether at least eight quite different innovations were combined to make the amniotic revolution possible: [1] the formation of a tough impervious shell; [2] the formation of

the gelatinous egg white (albumen) . . . [3] the excretion of nitrogenous waste . . . [4] the formation of the amniotic cavity in which the embryo floats; [5] the formation of . . . a container for waste products . . . [6] the development of a tooth or caruncle which the developed embryo can utilize to break out of the egg; [7] a quantity of yolk sufficient for the needs of the embryo till hatching; [8] changes in the urogenital system of the female permitting fertilization of the egg before the hardening of the shell.[43]

What's more, these complex innovations needed to develop simultaneously. What if only one of these innovations fails to manifest? What if everything evolved on time except for, say, the tooth the embryo needs to break out of the shell? Or the waste elimination mechanism necessary to prevent poisoning of the embryo? The answer, of course, is that the amniotic egg would fail to produce offspring, and natural selection would weed it (and hence reptiles) out.

These are but eight of many adaptations needed to transform an amphibian into a reptile. One might suppose, incorrectly, these many adaptations would be evidenced by many intermediate fossils.

Evolution or not, Planet Earth was at one time the home to a number of superbly impressive reptiles. The foul-tempered T-rex and his fellow dinosaurs come to mind. Dominant at one time, these bad boys of the Jurassic period died out well before the appearance of man.

Dinosaurs are portrayed as a testimony to evolution, but that is difficult to rationalize. Like almost everything else in the fossil record, dinosaurs suddenly appeared. Then they died out as rapidly as they came, with no indication that they served as a progenitor to new species.

As well as any cat burglar there ever was, evolution has an uncanny ability to cover its tracks.

~ 25 ~

REPTILES TO BIRDS

Sceathers and Lings

If evolutionists are correct, reptiles win the prize for diversity in progeny—they gave rise to both birds and mammals.

$$\text{fish} \rightarrow \text{amphibian} \rightarrow \text{reptile} \rightarrow \begin{array}{l} \text{bird} \\ \text{mammal} \end{array}$$

Indeed, some fossils have both reptile and bird features, and some have reptile and mammal features.

The most famous fossils that clearly show bird and reptile characteristics are those of *Archaeopteryx* ("ancient wing"). The first *Archaeopteryx* fossil (there are several in existence) was discovered in 1861, two years after Darwin published *The Origin of Species*. The specimen impressed Darwin.

And it continues to impress. The National Academy of Science encourages teachers to hail *Archaeopteryx* as a bona fide missing link that bridges the dinosaur-bird (walker-flyer) chasm.[44]

The Facts

Archaeopteryx had bird features, including:

- hollow bones (reptiles have heavy solid bones)
- modern bird feathers (with no intermediate scale-feather transitional evidence)
- wings designed for flight (asymmetrical rather than symmetrical feather arrangement)

and it had reptilian features, including:

- teeth (modern birds do not have teeth)
- long, boney tail
- claws on its wings

Archaeopteryx, or Archae for short, is typically presented as a feathered dinosaur-bird that could glide but not fly, as one might reasonably expect of a half-bird, half-reptile creature. When the reptile first got its wings, it hadn't evolved to the point of being able to flap them into flight, so apparently it did the next best thing: it took to gliding.

The Kitty Hawk Moment

One big debate centers on the creature's Kitty Hawk moment. Did flight originate from a four-legged creature that climbed a tree, jumped, and found it possessed the remarkable ability to glide (tree down), or from a creature that experienced lift while chasing, on its hind legs, its next meal (ground up)? The ground up theory seems to be favored at the moment. If true, natural selection must have been asleep at the wheel to allow a two-legged creature that survives by outrunning its prey evolve arm-wing structures susceptible to wind drag.

The Teeth

Much is made of the fact that Archae had teeth, whereas modern birds do not. I suppose Archae couldn't really have been *all* bird, given it had teeth. And since reptiles do have teeth, it must have been at least part reptile. Then again, turtles are reptiles that don't have teeth. Hmmm, that's curious. And some fish have teeth, and some do not. Some amphibians have teeth, and some do not. Some mammals have teeth, and some do not. I think I see a trend: different animal groups have both toothy and toothless representatives. Maybe it's best we hold off a while on making too big of a fuss over Archae's teeth.

The Timeline

We should be able to draw conclusions based on Archae's geological timeline. After all, the point of the story is that dinosaurs predate *Archaeopteryx*, and *Archaeopteryx* predates birds. One cannot simply note that Archae has features common to dinosaurs and birds: the integrity of the timeline is critical.

To illustrate the value of a timeline, pretend you are a detective investigating last week's armed robbery of the local bank. Suspect No. 2, whom we'll call Archie, looks particularly interesting, in light of the facts:

- witnesses described someone with features similar to Archie as the gunman

- a gun registered to Archie was found at the scene
- Archie's fingerprints were on the gun
- bullets accidentally discharged at the scene were shown by forensic specialists to be consistent with the make and model of bullets found at Archie's home
- a car registered to Archie was discovered to have been the getaway car

An open and shut case? Not so fast.

Admittedly, it looks very bad for Suspect No. 2. But things brighten considerably when Archie's lawyer, after placing all the evidence in a timeline, discovers Archie was in the county jail when the robbery occurred! The robbery might have been committed by someone who resembles Archie, driving Archie's car, and using Archie's gun (with fingerprints from the last time Archie used it), but the jury must acquit because the timeline doesn't fit. Despite circumstantial evidence, the defense attorney is able to falsify the prosecutor's conclusions with a timeline. (The robber turned out to be Archie's first cousin, which explains the circumstantial evidence.)

So where does Archae, who possesses both reptile and bird features, fit into the critical dinosaur-bird timeline? *Archaeopteryx* was discovered in Upper Jurassic sediment, and so is believed to have lived about 150 million years ago. However, in 1986 a 225-million-year-old bird fossil, *Protoavis texensis*, was discovered, making it 75 million years *older* than Archae. And since *Archaeopteryx* does not predates birds, it cannot be a missing link between dinosaurs and birds. To wreck the dinosaur-to-bird timeline further, the bird *Protoavis* dates almost as old as the oldest dinosaur, throwing the entire dinosaur–bird timeline into question.

Yet Another Mosaic

If Archae is not a missing link, then what is it?

> *Archaeopteryx* is a "mosaic" of useful and functioning structures found also in other creatures, not a "transition" between them. A true transitional structure would be, say, a "sceather"—that is, half-scale, half-feather—or a "ling"—half-leg, half-wing. . . .[45]

Since Archae had well-defined feathers (birds fall into class Aves, characterized by the presence of feathers), *Archaeopteryx* may have been an ancient bird with reptile features. All parts, including teeth, were fully functional.

Evidence for Archae has unraveled to the point that some paleontologists now doubt it is even a direct ancestor of modern birds.[46]

With *Archaeopteryx* losing its importance status as dinosaur-bird missing link, evolutionists needed a paleontological break, which they got one in the form of a fossil from China.

Archaeoraptor

In the 1990s an amazing fossil was discovered and subsequently smuggled out of China. The fossil revealed a 125-million-year-old creature with a bona fide dinosaur tail and feather-like features on its forelimbs. Christened *Archaeoraptor*, the missing link fossil was purchased for $80,000 and subsequently hailed in a *National Geographic* article entitled "Feathers for T. Rex?"[47]

Only one problem—but a very big one where fossils are concerned: it was a fake! A dinosaur tail had been glued to ancient bird parts.

Fortunately, there is a thin silver lining to this story. Unlike the Haeckel situation which lasted over than a century, scientist decided to reject this fraud in relative short order.

With *Archaeoraptor* a counterfeit and *Archaeopteryx* an untimely mosaic, and the gap between reptiles and birds remains as wide as ever. Sceathers and lings remain elusive.

REPTILES TO MAMMALS

I Object, Your Honor.

For reptiles to have given rise to mammals, a number of interesting transitions must have occurred. Among other things, reptiles needed to:

- drop their scales and grow hair
- convert their multi-bone jaw into a single one
- convert their ear from one with a single bone to one with three bones (the so-called hammer, anvil, and stirrup)
- convert from cold-blooded (body temperature controlled by the environment) to warm-blooded (body temperature internally regulated by burning fat)
- grow mammary glands

In light of irreducible complexity, the accomplishment of any one of these changes by blind chance is diminishing small, and the notion that all five could occur is nothing short of biological fantasia. To be sure, hundreds of publicly funded biology departments from across the globe generate mountains of data every year, a source rummaging speculators use to piece together just-so stories. The volume and technical nature of the research places it beyond the reach of the average non-biologist, so the entire field of Darwinian evolution is a sticky wicket, indeed.

But as a jury of non-lawyers uses its collective common sense and reasoning to come to a verdict following detailed legal arguments, so non-biologists can use the same skill sets to draw conclusions from esoteric arguments based in the biological sciences.

The Case Made

Members of the jury, I now state my case: reptiles evolved into mammals. Appreciating the esoteric nature of the raw data, and being sympathetic to the fact you are non-technical in your formal training, I respectfully state my case in basic terms. But please remember that I have a mountain of technical data to support my claims, which appear strange—even bizarre—at times.

Let us begin with differences between reptilian and mammalian ears. A reptile has three bones in its jaw and one in its ear, whereas a mammal has one bone in its jaw and three in its ear, as the following exhibit shows.

	Number of Jaw Bones	Number of Ear Bones
Reptile	3	1
Mammal	1	3

It is easy to see what happened. In the production of a mammal from a reptile, two bones of the reptilian jaw shifted to make a three-boned mammalian ear. This is how the mammalian ear came to have three bones. Also note the loss of two bones from the reptilian jaw leaves behind a single jawbone, which is exactly what we observe in modern day mammals. These common sense observations, I assure you, are backed by the opinions of experts.

As to details, the two migrating jawbones evolved into the middle ear bones you recognize as the *hammer* and the *anvil*. They slid in between the eardrum and what is now referred to as the *stirrup* . . . with no loss of hearing to the evolving species. Of course, the jaw also remained operable throughout the transition. Scientists don't know exactly how ears and jaws remained operable during the drastic transition, but this is of no great concern—just because someone doesn't know *how* something occurred doesn't mean that it didn't occur. All scientists in good standing with their trade appreciate this axiom.

To continue, modification, migration, and alignment were precise to the point the three bones could communicate through tapping, and pass that tapping along to the inner ear. Thus, a daisy chain of communication was established between atmospheric air compressions and electrical impulses to the brain. As one would expect, natural selection maintained such an advantageous innovation of evolution.

Of course, I need a little leeway in my scenario, as the details may need to be tweaked as new data comes in. For example, if a jawbone disappeared with the concomitant arrival of a new ear bone (rather than the same bone actually migrating), the net result would stand.

This story may sound incredible to the novice, but don't forget that given enough time anything is possible. You have the evidence before you—as sure as these words ring true in your own ears!

Counter-Argument

Ladies and gentlemen of the jury, let me confirm there is indeed a mountain of technical biological data, but precious little provides support for the wildly speculative scenario of bone modification, migration, and integration, with no loss of hearing or jaw functionality.

In the previous statement, you heard part of the reptile-to-mammal ear story—but there's much more in play than a couple of migratory bones. For example, unlike reptiles, mammals also have a very complex structure in their inner ear called the *organ of Corti*. As to how such an organ, specialized as the retina of the eye, could have evolved from a reptilian ancestor is a mystery, given reptiles do not have corresponding organs. It's one thing to suppose existing body parts shift around, it's quite another to create them out of thin air.

I appreciate, as do those on the other side of this argument, that the mammalian ear is a very complex biological machine. In fact, complex to the point of being considered *irreducibly complex*—which of course, removes from possibility a scenario in which bones gradually integrate into the middle ear, and fully functional organs appear on queue.

And to suppose two jawbones could squeeze in between the eardrum and stirrup with no loss of hearing is, by itself, beyond incredible. But if there were a loss of hearing, then there was no advantage; and if there was no advantage, then natural selection would have weeded out the mutant.

As a general observation, it is curious that there are no examples of reptile descendents with two ear bones. All reptiles have one ear bone, and all mammals three. As far as a smooth transition, Darwinian evolution seems to have skipped a beat when it came to shifting ear bones.

Let me close with a couple of questions. (1) Have you ever observed anything as complex as, say, a lawn mower come about by chance? (2) Which do you consider more complex: your lawn mower or your ear?"

And On It Could Go

Of course, the exchange could linger for months. The details (or lack thereof) of the cold-blooded-to-warm-blooded transition need to be addressed, as well as the development of hair. Did hair evolved from one of the ancestors of lizards (proteins similar to ones used to make claws are found in hair), or did it evolve directly from scales?

And of course, there's the evolution of mammary glands and milk production. Maybe mammary glands developed from sweat glands. Maybe milk, a complex cocktail of proteins, sugars, and fats, is supercharged sweat (or some other secretion). Maybe some of its proteins evolved from existing proteins, while other proteins were co-opted from sources such as blood. Once together, ostensibly for the purpose immunoprotection against invaders, maybe proteins discovered they could process sugars and fats and provide a yummy meal for infants. What a great innovation of evolution: the ability to feed and immunize offspring simultaneously. And maybe through inflammation, the gland grew into a breast.

And maybe toy boats really do evolve into Mississippi steamships.

~ 27 ~

LAND MAMMALS TO SEA MAMMALS

Another Fishy Story

In an amazing twist, at least one mammal decided to return to the water from whence its ancestors came. Fortunately, the extreme flexibility of Darwin's theory allows it to accommodate any ancestor-descendant path imaginable, including U-turns. To wit, the story of the whale, which extends back to the musings of Mr. Darwin himself:

> In North America the black bear was seen by Hearne swimming for hours with widely open mouth, thus catching, like a whale, insects in the water. Even in so extreme a case as this, if the supply of insects were constant, and if better adapted competitors did not already exist in the country, I can see no difficulty in a race of bears being rendered, by natural selection, more and more aquatic in their structure

and habits, with larger and larger mouths, till a creature was produced as monstrous as a whale.[48]

Apparently, the black-bear-eating-insects story didn't work out so well. But undaunted and in search of facts to support a conclusion, theorists came up with another terrestrial mammal-to-aquatic mammal storyline. A number of fossils were arranged in a lineage of sort, starting with a hyena-like mammal and ending with a whale. The series takes shape as spotty information is cast in the form of "artist's rendition" type drawings. The tail of the mammal morphs into a pair of horizontal tail fins, front legs morph into flippers, rear legs disappear, eyes adjust for underwater sight, eardrums adjust for high pressures, nostrils become blow-holes, and mammal body parts that can't be matched up with whale counterparts become vestigial organs.

Besides the astronomical (im)probabilities associated with evolving the myriad of irreducibly complex mechanisms necessary to move across the hyena-to-whale series, conclusions among experts based on morphological and molecular evidence have not been reconciled.[49]

From a candid interview with an official of the American Museum of Natural History in which the subject of whales comes up, interviewer Luther Sunderland records the museum representative as having remarked that scenarios of whale evolution are limited only by a scientist's "own imagination, and the credulity [gullibility] of the audience."[50]

Caveat emptor.

~ 28 ~

MAMMAL TO MAN

Into the Boneyard

Okay, time for the great debate: monkey-to-man.

Actually, the monkey-to-man debate is overemphasized. Filling the gap between an ape-like ancestor and man is less significant than filling in the missing links between fish and amphibians, or solving the mystery of the Cambrian Explosion.

The reason for the disproportionate interest is the closeness with which the monkey-to-man debate hits home. Cambrian creatures are one thing, but a hairy biped that walks upright and looks a little like Uncle Bob is quite another.

Skeletons in the Closet

If the public has its doubts about the evolution of man from ape-like ancestors, paleoanthropologists (those who study human beings from ape- and human-like fossils) should be tolerant of that sentiment. After all, the public has been especially tolerant of misinformation from the field of paleoanthropology.

In December 1912, Charles Dawson, honorary collector for the British Museum, announced the discovery of the Piltdown Man, an ape-man missing link. Piltdown Man was rationalized from a man-like skull, an ape-like jawbone, and a tooth intermediate to man and ape. The Piltdown Man was hailed as a bona fide missing link.

For decades Pilty served as poster boy for evolutionists, despite quiet suspicions of authenticity. The paleoanthropologist community at large supported the Piltdown find, which generated hundreds of written papers.

However, four decades or so after the original discovery, the heralded missing link was proved to be a fraud:

- Radiocarbon dating determined the Piltdown skull to that of a human.
- Collagen reactions showed the jaw to be that of an orangutan.
- Chemicals were used to give the jaw the appearance of age.
- File markings indicated the teeth had been filed down to make the teeth look more human-like.

In support of a darling theory, fudged and manipulated data was thrust on an unsuspecting and trusting (or as others might put it, gullible) public, while experts somehow missed the obvious.

In his book *Tornado in a Junkyard*, James Perloff summarizes a few notable misfires in the field of paleontology:

> By the 1950's, human paleontology was suffering from a bad case of the blahs. Piltdown Man

had been exposed as a fraud; Nebraska Man was a pig's tooth; Java Man's discoverer had called him a genus allied to the gibbons; Peking Man's fossils had flown the coop; and Neanderthals had turned out to be Homo Sapiens.[51]

Into the Boneyard

By the mid 1970s, the fossilized remains of nearly four thousand *hominids* (humans plus evolutionary ancestors) had been stockpiled in various scientific vaults. With the plethora of skulls and bones came a number of arguments: Is his fossil older than her fossil? Is this skull a *Homo sapien* or a Neanderthal? Is that skull a *Homo erectus* or an Australopithecinae?

The exact chronology is difficult for the layman to pinpoint. Paleoanthropologists might rationalize ape-men, old-world monkeys, Australopithecus, and other creatures differently, but they are virtually unanimous in one thing: man evolved from something with a sloped forehead that walked on its knuckles.

The lineage of man falls roughly in line with the following table. Some ape-like creature begat Australopithecinae (southern man-apes), which begat *Homo habilis*, which begat *Homo erectus*, which begat Neanderthal, which begat *Homo sapiens* (modern man).

Timing	Item	Classification
modern	1	Homo sapiens
	2	Neanderthal
	3	Homo erectus
	4	Homo habilis
	5	Australopithecinae
oldest	6	Ape-like ancestor

For twenty-five years Marvin Lubenow slogged through the boneyards to deliver an insightful book entitled *Bones of Contention*. He comes to some straightforward conclusions regarding the ancestry of man. The following five quotes (brackets are mine) help to clarify what is man and what is monkey.

> The facts of the big picture are that first, fossils that are indistinguishable from modern humans can be traced all the way back to 4.5 m.y.a., according to evolution time scale. That suggests that true humans were on the scene before the australopithecines [Item 5] appear in the fossil record. [Hence, true humans could not have evolved from the australopithecines.]

> Second . . . [t]he fossil record does not show [*Homo*] *erectus* evolving from something else or evolving into something else.

> Third, anatomically modern *Homo sapiens*, Neandertal, archaic *Homo sapiens*, and *Homo*

erectus [Items 1, 2, and 3] all lived as contemporaries at one time or another.

Fourth, all of the fossils ascribed to the *Homo habilis* [Item 4] category are contemporary with *Homo erectus* [Item 3]. Thus, *Homo habilis* not only *did* not evolve into *Homo erectus*, it *could* not have evolved into Homo erectus.

Fifth, there are no fossils of *Australopithecus* or of any other primate stock in the proper time period to serve as evolutionary ancestors to humans. As far as we can tell from the fossil record, when humans first appear in the fossil record they are already human.[52]

Lubenow sees no transitions from ape-like ancestors to modern humans. *Homo habilis, Homo erectus,* Neanderthal, and *Homo sapiens* (Items 1 through 4) are human-like, whereas *Australopithecus* and its supposed ancestors (Items 5 and 6) are ape-like. Lubenow's conclusions are summarized in the following table, in which all specimen fall into one of two distinct groups: human-like and ape-like.

Group	*Classification*
human-like	Homo sapiens
human-like	Neanderthal
human-like	Homo erectus
human-like	Homo habilis
ape-like	Australopithecinae
ape-like	Ape-like ancestor

More recently, an article in the online journal of the International Society for Complexity, Information, and Design (ISCID) remarks:

> *Homo* [abruptly appeared] as a novel and distinct form, significantly different from earlier fossil forms and without links to previous fossil forms.... *Homo* is proposed as a basic type, with current members of *Australopithecus* plus what is currently labeled *Homo habilis* suggested as another extinct basic type. The species remaining within *Homo* have similar morphologies that can generally be explained as microevolution within a basic type.[53]

Here, the classification of all specimen into two groups is confirmed, only the dividing line is slightly shifted to include current members of *Homo habilis* into the ape-like category.

As with fish-to-amphibian, amphibian-to-reptile, and reptile-to-mammal transitions, the alleged monkey-to-man transition offers yet another unfounded chronology.

Missing links for all geological periods comprise a supreme embarrassment evolutionists would love to be able to explain. Holes in the fossil record are as legion as they are undeniable.

> *I did not say . . . that the fossil record contains*
> *no intermediate forms; that is silly. What I did*
> *say was that there are gaps in the fossil graveyard,*
> *places where there should be intermediate forms but*
> *where there is nothing whatsoever instead. . . .*
> *Darwin's theory and the fossil record are in*
> *conflict. . . . nothing is to be gained by suggesting*
> *that what is a fact in plain sight is nothing of the*
> *sort.*[54]
>
> —DAVID BERLINSKI, *Uncommon Dissent*

What evolution needs is a new adjunct theory to resuscitate dying Darwinism.

And right on queue, the champions of evolution concocted one . . .

~ 29 ~

PUNCTUATED EQUILIBRIUM

The Best Evidence There Never Was

One might reasonably anticipate that millions of years of evolution would leave a fossil record replete with intermediate forms. (Darwin thought so.) Instead, the fossil record fails to validate evolution. Few, if any, examples of smooth progressions from less to more complex species are to be found. (Darwin was wrong.)

Evolutionists are divided as to how to handle this situation. Some use the *sour grapes approach*: they ignore or greatly downplay the importance of the lack of transitional forms. "Sure, the missing links are still missing, but our theory doesn't need them anyway." Others use the *state of denial approach*: "Gaps? What gaps? I don't see any gaps." But some bravely admit the fact that the fossil record is riddled with holes, and look for a new explanation.

Enter *punctuated equilibria*, a relatively new evolutionary model that attributes the glaring absence of intermediate forms between species to occasional bouts of rapid evolutionary activity. So, it appears that evolution has two gears in its transmission case: *punctuated equilibria* is high gear; and Darwinian *gradualism* is low gear.

Punctuated equilibria is a spin-off of the older and much ridiculed *Hopeful Monster model*, which is based on the idea that a severely deformed offspring (essentially a monster) could be birthed with characteristics that allow it to survive and serve as a new species. Imagine a reptile laying an egg and a fully functional bird hatching from it. To a scaly reptile, a feathered bird would indeed have seemed a monster (albeit an extremely impressive one). In such a case, one would not expect to find an intermediate form between reptile and bird. As an aside, it is interesting to ponder how an animal (in this case, a bird) created by such a rapid process could reproduce. If it were the only one, with what would it mate?

Punctuated equilibria is more sophisticated than the Hopeful Monster model and doesn't claim transitions as large as reptile-to-bird. Loosely speaking, it claims there are long, quiet periods of time during which little or no evolution occurs (*equilibrium*), and there are spurts of evolution that occasionally disrupt (*punctuate*) the equilibrium. (A rapid bout of evolution may entail tens of thousands of years, as opposed to millions.)

Let's think through the concept of punctuated equilibria. During *equilibrium*, little or no evolution occurs, so missing-link fossils are not left behind. During *punctuation*, evolution occurs so rapidly that missing-link fossils are not left behind.

How convenient! Evolution occurs too slowly to fill in the evolutionary gaps, or it occurs too rapidly to fill in the evolutionary gaps. Marvin Lubenow puts the situation in proper perspective:

> Certainly, the punctuated equilibria theory is unique. It must be the only theory ever put

forth in the history of science which claims to
be scientific but then explains why evidence for
it cannot be found.[55]

Think of Punctuated Equilibria as *Hopeful Monster Lite.*
It claims the same net result, and it is burdened by the same
insurmountable obstacles. There is no proof that it ever has
or could work, and it runs headlong into the buzz-saw of irreducible complexity.

The evidence pool suggests Darwinian evolution has
never been in high gear or low gear: it has always been in
park.

~ 30 ~

MOLECULAR BIOLOGY

*What Do Yeast, Tuna, Pigeons, and Horses
Have in Common?*

Evolutionists welcomed the advent of molecular biology with open arms, given the unpersuasive fossil record. The Cambrian Explosion set the tone for the entire fossil record: all major types of organisms arrive fully formed and functional.

Then again, fossil record data *is* qualitative rather than quantitative. It is "soft" data. Conclusions are drawn largely from opinions and consensus, rather than measurable benchmarks.

Maybe that's the problem—maybe it is too soft. Maybe support for Darwinism really is buried somewhere in the data, it just gets lost in all the opinions, counter-opinions, *ad hominem* attacks, and just-so stories. My bird with teeth may be your reptile with wings; my fish with hair may be your mammal with flippers; yesterday's fish-amphibian intermediate may be tomorrow's fish.

Hurrah! Hard Times Are Here

The discovery of DNA ushered in molecular biology, a new and exciting area of study that promised to provide "hard" data—the kind of data not burdened by the baggage of subjectivity.

Molecular biology focuses on biomolecules, complex compounds that run our biological processes and store our genetic information. Molecules such as DNA (which contain an organism's genetic information and provide the chemical blueprint for the constructing new organisms) and proteins (chemical workhorses whose compositions, shapes, and activities are predetermined by the DNA blueprint) are prime targets of investigation.

Biomolecules can be analyzed and compared, providing an indication of variation among species. Analyses are accurate and quantitative. Opinions may differ as to how to interpret the data, but not to the solidity of the data itself. Guesswork is kept to a minimum.

In his important book *Evolution: A Theory in Crisis*, molecular biologist and medical doctor Michael Denton looks at evolution's single-cell-organism-to-man scenario through the prism of biochemical data.[56,57]

one-cell organism → fish → amphibian → reptile → bird / mammal

A couple of the molecular biology experiments will be highlighted; but first, some necessary background.

Cytochrome C: A Benchmark for Variation

One particularly interesting biomolecule common to all members of the preceding sequence is *cytochrome C*, a biochemical that varies form species to species. As Yogi Berra might put it, "It's the same, only different." It's the same molecule, only it's different from species to species.

It is variation among species that makes biomolecules such as cytochrome C valuable as a research tool. The more a species differs from a supposed ancestor, the more its biomolecules should differ from the biomolecules of that ancestor. For example, if bacteria are ancestors of fish, and if fish are ancestors of humans, then bacteria cytochrome C should

bacteria ➝ fish ➝ human

resemble fish cytochrome C more than it does human cytochrome C.

How Is Variation Measured?

To illustrate how *variation* among cytochrome C molecules is determined, consider the following three sequences of ten letters each:

> A-B-C-D-E-F-G-H-I-J (Sequence 1)
> A-B-<u>B</u>-D E-F-G-H-I-J (Sequence 2)
> A-B-<u>B</u>-<u>B</u>-E-F-G-H-I-<u>S</u> (Sequence 3)

Sequences 1 and 2 differ at the third digit only (Sequence 1 has a "C" whereas Sequence 2 has a "B"). Since they differ by one digit out of ten, there is a ten percent sequence difference. Similarly, Sequences 1 and 3 differ at the third, fourth, and tenth digits. Because they differ by three out of ten digits, they have a thirty percent sequence difference. Therefore, Sequence 1 is closer to Sequence 2 (from which it varies by ten percent difference) than it is to Sequence 3 (from which it varies by thirty percent).

Sequence	Percent Difference Compared to Sequence 1
2	10 %
3	30 %

Percent sequence differences are similarly calculated for proteins, which are sequences of amino acids rather than letters. Cytochrome C is composed of 104 amino acids units, so it would make a rather long word by analogy.

Experiment 1
Cytochrome C: From Prokaryote to Eukaryote

In the first illustrative experiment, percent sequence differences between cytochrome C for bacteria and cytochrome C for each of six representative organisms (yeast, wheat, silk moth, tuna, pigeon, and horse) are determined.

This is a particularly interesting study for couple of reasons. Firstly, bacteria represent the simplest self-replicating cell. As far as we can tell, bacteria are ground zero (or very nearly so) for the formation of life from non-living materials. Fortunately, species ranging from bacteria to man are in existence today, so evolution can be studied, via molecular biology, on readily available organisms. Should a molecular biologist have concerns about the structure of the species supplying the cytochrome C, she has merely to cut one open and take a look.

Secondly, the structure of bacteria cells is markedly different from those of the six organisms to which it is to be compared. Bacteria are *prokaryotes*, which means their cells do not have a nucleus; whereas yeast, wheat, silk moths, tuna, pigeons, and horses are *eukaryotes*, meaning their cells have nuclei. Hence, this experiment studies organisms whose cell plans differ at the highest levels.

The results are stunning. As shown in the following table, the difference between bacteria (prokaryote) cytochrome C and cytochrome C from each of the six comparison organisms (eukaryotes) is almost the same!

Organism	Class	Percent Difference Compared to Bacterial (Prokaryote) Cytochrome C
yeast	eukaryote	69 %
wheat	eukaryote	66 %
silk moth	eukaryote	65 %
tuna	eukaryote	65 %
pigeon	eukaryote	64 %
horse	eukaryote	64 %

Bacterial cytochrome C is no closer to wheat cytochrome C than it is to horse cytochrome C. Although the six eukaryotes differ among themselves, all are nearly the same "distance" from the benchmark prokaryote (bacteria):[58]

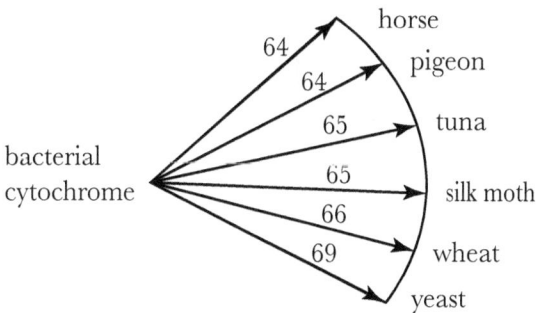

Cytochrome C Percent Sequene Difference
(procaryote vs. eucaryote)

Denton concludes, "The bacterial kingdom has no neighbour in any of the fantastically diverse eucaryotic types. The 'missing links' are well and truly missing."[59]

Experiment 2
Cytochrome C: From Fish to Fowl

The second experiment of interest uses cytochrome C from representative animals to test Darwin's descent sequence:

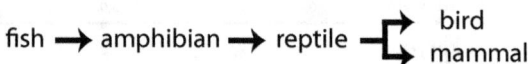

fish → amphibian → reptile → bird / mammal

Test subjects include a fish (carp), an amphibian (bullfrog), a reptile (turtle), a bird (chicken) and two mammals (rabbit and horse). This sample set allows us to compare five different types of land vertebrates, some of which supposedly evolved long before others, against an aquatic vertebrate (carp).

Results are as blackening to Darwinism as those from the first experiment. Percent differences between carp and each of the land vertebrates are nearly identical—about 13 percent.[60]

Organism	*Class*	*Percent Difference Compared to Carp (Fish) Cytochrome C*
bullfrog	amphibian	13 %
turtle	reptile	13 %
chicken	bird	14 %
rabbit	mammal	13 %
horse	mammal	13 %

The trend (or lack thereof) is easy to see: a carp (fish) is no closer to a bullfrog (amphibian) than it is to a turtle (reptile), a chicken (bird), or a horse (mammals). These and many other molecular analyses draw to the same conclusion as the fossil record: the data is inconsistent with a continuous spectrum of slightly different creatures across time, one melding into another, and then another, ultimately producing more complex organisms from less complex organisms. Rather, organisms tend to fall into distinct and unbridged groups.

The idea of common descent foiled yet again.

PART FIVE
The Long Journey Done

~ 31 ~

CHARLES DARWIN PREDICTS

He Caught The Beagle, but Missed The Boat

In addition to being falsifiable, scientific theories are predictive. They make bold claims, as if to say, "Here's what I predict, and I dare you to prove me wrong."

Charles Darwin, famed naturalist who sailed *The Beagle* into scientific lore, observed microevolution (horizontal evolution) and theorized macroevolution (vertical evolution). Being a bright chap, he knew the value of accurate predictions to a scientific theory, so he made some.

Darwin's most daring predictions were made possible by the incompleteness of the fossil record during his time. He used his theory to predict what the fossil record would eventually show once fossil hunters had unearthed enough fossils to render the fossil record functionally complete.

> If my theory be true, numberless intermediate
> varieties . . . must assuredly have existed . . .
> [and] evidence of their former existence could

> be found only among fossil remains ... [which
> is] in an extremely imperfect and intermittent
> record.

Here, Darwin suggests that "numberless intermediate varieties" will eventually be found once the fossil record is no
longer "imperfect," thus eliminating what we now refer to as
missing links.

For assurance that Mr. Darwin has not been taken out
of context, the full paragraph from *The Origin of Species* is
provided (emphasis mine):

> Lastly, looking not to any one time, but at all
> time, *if my theory be true, numberless intermediate va-
> rieties*, linking closely together all the species of
> the same group, *must assuredly have existed*; but
> the very process of natural selection constantly
> tends, as has been so often remarked, to exter-
> minate the parent forms and the intermediate
> links. Consequently *evidence of their former exis-
> tence could be found only among fossil remains,* which
> are preserved, as we shall attempt to show in a
> future chapter, *in an extremely imperfect and inter-
> mittent record.*[61]

Guess what? The fossil record is no longer extremely
imperfect. In fact, it is not even kind-of-sort-of imperfect.
There are now millions upon millions of fossils, with abundance from every geological time period. It's in great shape.
Our archaeological bucket runneth over.

Guess what else? Numberless intermediate varieties have
not been found. The intermediate links between groups—
they are still missing. Of the ones proposed, all are burdened

by numerous irreducibly complex structures between nearest neighbors.

Unlike today's evolutionists who treat the fragile theory with kid gloves, Charles Darwin made gutsy predictions. More than once he claimed that great numbers of transitional creatures must have lived:

> [T]he number of intermediate and transitional links, between all living and extinct species, must have been inconceivably great. But assuredly, if this theory be true, such have lived upon the earth.[62]

And more than once he blamed the incompleteness of the fossil record for missing links:

> Why then is not every geological formation and every stratum full of such intermediate links? Geology assuredly does not reveal any such finely graduated organic chain; and this, perhaps, is the most obvious and serious objection which can be urged against my theory. The explanation lies, as I believe, in the extreme imperfection of the geological record.[63]

and

> Now let us turn to our richest museums, and what a paltry display we behold! That our collections are imperfect is admitted by every one.[64]

and

I believe the answer mainly lies in the record being incomparably less perfect than is generally supposed.[65]

But even in light of a limited fossil record, the obvious lack of support for his position was "difficult to understand":

[I]t cannot be doubted that the geological record, viewed as a whole, is extremely imperfect; but if we confine our attention to any one formation, it becomes much more difficult to understand why we do not therein find closely graduated varieties between the allied species which lived at its commencement and at its close.[66]

The reason "why we do not therein find closely graduated varieties between the allied species" may not be as perplexing to some as it was to Darwin. Perhaps, we don't find them because they never existed.

Darwinian evolution has been falsified—at least, by the challenges laid down by its own namesake. Every gap in the modern-day fossil record screams that Darwin was wrong, and every vote for Punctuated Equilibria (which substantiates the lack of transitional fossils) echoes that scream.

Darwin might have caught *The Beagle* to Galapagos and correctly assessed the role of natural selection in microevolution, but he missed the boat regarding macroevolution and the origin of species.

~ 32 ~

THE FLIP SIDE OF EVOLUTION

The Spirit of Mr. Mivart Lives

And so we come to the end of our twenty-billion-year trek. Along the way we have seen evolution claim to . . .

- form a universe from nothing
- synthesize by chance the macromolecules necessary for life
- organize nonliving materials into living creatures
- simultaneously produce fully functional members of every animal phylum
- convert lower creatures into fish
- transmogrify fish into amphibians
- transmute amphibians into reptiles
- metamorphose reptiles into birds and mammals
- transfigure ape-like creatures into a self-actualizing mammal called "man."

It has been said that evolution is a fairy tale for adults. Given that evolution touts the ability to turn frogs into princes, the claim may not be far off target. No kiss from a princess is necessary—a few eons of time will do the trick.

In truth, there is not enough time in all of history to overcome the insurmountable odds of forming, by chance, life from the nonliving; and no amount of time, including eternity future, will allow evolution to circumvent the laws of thermodynamics or unfavorable mathematical probabilities. "Favorable mutations" is an oxymoron of science. The gappy fossil record argues against Darwinism. And Punctuated Equilibria, which highlights the problem of missing links, dies instantly in the face of irreducible complexity.

> *We're still waiting for Darwin's Newton: for a theorist who can take Darwin's proposal and produce even one hypothesis about the origin of one interesting biological mechanism, a hypothesis which specifies, step by step, the genetic changes that had to take place, the embryological alterations that these changes produce, and the quantifiable selective pressures that enable each new step to reach a significant proportion of the population.*[67]
> —ROBERT KOONS, *Uncommon Dissent*

Evolution: The Ultimate Protected Species

Evolution survives because it is exempted from rigorous scientific scrutiny. Debates within the scientific establishment center around issues of *how* evolution occurs, not *whether* it occurs. The length and style of the king's clothes may be up for debate, but the issue of whether he is actually wearing clothes is not.

> *So far we have seen that there is ample reason to believe that Darwinism is sustained not by an impartial interpretation of the evidence but by dogmatic adherence to a philosophy even in the teeth of evidence.*[68]
> —PHILLIP E. JOHNSON, *Defeating Darwinism*

Where education and scientific debate are concerned, the trouble is not that claims favorable to evolution are taught, but that claims unfavorable to the theory are censored. Science is a self-regulating endeavor. Those on both sides of any issue should have their say. Such is not the case: Darwinists have near exclusive access to the academic podium. The fix is in.

Ironically, once upon a time Darwinists were the champions for equal access. Charles Darwin cried foul when a staunch critic of evolution, Mr. Mivart, appeared to present only one side of the story—the side that opposed evolution. Darwin complained:

> [A]s it forms no part of Mr. Mivart's [Darwin's critic] plan to give the various facts and considerations opposed to his conclusions, no slight effort of reason and memory is left to the reader, who may wish to weigh the evidence on both sides.[69]

The situation has turned 180 degrees since Darwin's time: the victim has become the perpetrator. Darwinism now reigns supreme in academia, but it does so in the spirit of

Mr. Mivart, not Mr. Darwin. It is now the Darwinists who do not want others to "weigh the evidence on both sides."

Evolution and ID

Despite its popularity among academic conformists, evolution remains a theory in crisis, and things are getting worse. Much worse.

A new revolution is turning science on its ear. Mentioned earlier, it is known as Intelligent Design (ID), a theory that runs counter-current to Darwinism. Darwinists claim biological systems arose by blind natural forces, despite the presence of "apparent design." ID theorists hold that design in biological systems is real and purposeful, and that design requires a designer. The proverbial battle line has been drawn in the sand.

It is important to note that ID theory is an endeavor limited to the detection of design; it does not address who or what the designer is. Maybe it's the Directed Panspermia aliens, maybe its Spielberg's "The Force," or maybe it's the creationist's God. "Is design present?" and "Who made the design?" are entirely different questions.

One thing is certain: if ID theory is correct, then Darwinian evolution is wrong. ID is on sound footing, and it succeeds where Darwinism fails. Darwinism, founded on the idea of creation through unguided natural forces, cannot rationalize the existence of irreducibly complex biological machines—ID Theory can. ID is also consistent with the observation that organisms fall into distinct groups; it anticipates a gappy fossil record.

> *Getting design without a designer is a good trick,*
> *indeed. Darwin was like a magician, performing*
> *far enough away from his subjects that he could*
> *dazzle them, until someone hands out binoculars.*
> *Darwin's idea was a good trick while it lasted, but*
> *with advances in technology as well as the*
> *information of the life sciences, especially*
> *molecular biology, the Darwinian magic gig is now*
> *up. It's time to lay aside the tricks, the smoke*
> *screens, and the hand waving, the "just so" stories*
> *and the stonewalling, the bluster and the bluffing,*
> *and to explain scientifically what people have*
> *known all along, namely, why you can't get design*
> *without a designer.*[70]
> —WILLIAM DEMBSKI, *The Design Revolution*

There is little doubt that more and more scientists will embrace the ideas of ID theory, but it will take some time. Paradigms are slow to change. Although many young scientists may not be willing to spend their careers supporting outdated ideas, Darwinism need not fear a scientific mutiny just yet. Old-school Darwinists who hold the power will cling to the theory of evolution as tightly as medieval intellectuals held to the Flat Earth theory.

In the meanwhile, academia will continue to provide Darwinism sanctuary. Students who may wish to "weigh the evidence on both sides" will continue to be, unbeknownst to them, denied the opportunity to do so without bias. ID theorists will continue to make progress and win converts. And dissenting scientists will strive to keep unearthed and within view . . . the flip side of evolution.

RESOURCES

Books and DVD's to Get You Started

Books

~ *TORNADO IN A JUNKYARD*, BY JAMES PERLOFF

Freelance writer James Perloff presents an excellent survey of evidences against evolution. Perloff is a non-scientist whose keen eye and sharp pen allow him to cover difficult subjects in layman's terms. This book provides a healthy dose of pictures and anecdotes, making it as entertaining as informative. I'd start here if I were you.

~ *ICONS OF EVOLUTION*, BY JONATHAN WELLS

Jonathan Wells moved to the top of the Darwinists' hit list when he published *Icons*. He dissects and simultaneously destroys the credibility of ten of evolutionists' flagship concepts, including: the famous Miller-Urey Experiment, in which life was "virtually created in a test tube;" Ernst Haeckel's fake embryo drawings; four-winged fruit flies . . . that don't, and everybody's favorite; *Archaeopteryx*, the former reptile-bird missing link that's been demoted out of the direct lineage of modern birds.

~ *Darwin's Enigma*, by Luther Sunderland

Sunderland recounts details of candid interviews "with officials in five natural history museums containing some of the largest fossil collections in the world." An older book, but not to be missed. If you ever doubted whether the public and private faces of Darwinism are the same, read this book and wonder no more. Prepare for many jaw-droppers as you read what the world's leading authorities say, and then compare those remarks to what you were taught in high school or college biology.

~ *The Politically Incorrect Guide to Darwinism and Intelligent Design*, by Jonathan Wells

Yes, it's the same Jonathan Wells who wrote *Icons of Evolution*, and this time he is in equally good form. He discusses what's wrong with Darwinism, and what's right with Intelligent Design. Along the way, he provides shocking anecdotes on the cut-throat business of science. The WWF has nothing on these guys—it's an ugly, no-holds-barred spectacle in which reputations are impugned and livelihoods are lost.

~ *EVOLUTION: A THEORY IN CRISIS*, BY MICHAEL DENTON

Since the 1980's this book has played a major role in the Evolution–Anti-evolution debate. It's a slightly tougher read than those books listed above. On the occasions Denton gets techy, but he provides background in layman's terms to help get you through. Do not overlook this book. Denton's book also provides a nice introduction to molecular biology (DNA, proteins, genes, etc.). Molecular biology is one of the newest battleground areas, and hence a topic those interested in the debate should become aware.

~ *DEFEATING DARWINISM*, BY PHILLIP JOHNSON

Phillip Johnson's "book aims to give . . . a good high-school education in how to think about evolution." He discusses many flaws of Darwinism, but more importantly, covers critical thinking skills the layman needs to credibly process and evaluate information regarding evolution. Johnson is a lawyer (but we won't hold that against him) who knows how to structure arguments, and he is a professor who knows how to relay information to those less knowledgeable. This book provides a good framework for the information you accumulate over time. If you wish Johnson had covered the scientific evidence in greater detail, well, he did—it's in his book, *Darwin on Trial*.

~ *DARWIN'S BLACK BOX*, BY MICHAEL BEHE

Behe presents the idea of irreducible complexity. This book, a "biochemical challenge to evolution," severed the Achilles' Heel of Darwinism. Don't panic over the word "biochemical"—Behe writes to be understood by non-scientists. Not only is the concept of irreducible complexity in and of itself important, it sets the stage for another important concept: Intelligent Design.

~ *The Design Revolution*, by William A. Dembski

If Darwinism isn't science's answer to the question of origins, what is? William Dembski's *The Design Revolution* provides the answer in clear and crisp terms. Each of the forty-four chapters is brief and may be read independently of the others. Skip around if you like, but read you must. An amazingly informative book. For those on the go, get the audio book.

~ *Darwin Strikes Back*, by Thomas Woodward

Thomas Woodward chronicles the rise of science's growing Intelligent Design revolution, despite the efforts of the Darwinian establishment. His book addresses flawed Darwinian arguments, not in isolation, but in view of ID. A very readable book which cuts to "the heart of the matter." It is endorsed by William Dembski, Phillip E. Johnson, Michael Behe, and Chuck Colson, among others. I paid $15 for my copy—it's worth triple the price.

DVD's

~ *ICONS OF EVOLUTION*

Based on Jonathan Well's book *Icons of Evolution*
"The Growing Scientific Controversy over Darwin"
Produced by ColdWater Media
www.coldwatermedia.com

~ *UNLOCKING THE MYSTERY OF LIFE*

"The Scientific Case for Intelligent Design"
Produced by Illustra Media
www.illustramedia.com

~ *DARWIN'S DILEMMA*

"The Mystery of the Cambrian Fossil Record"
Produced by Illustra Media
www.illustramedia.com

~ *WHERE DOES THE EVIDENCE LEAD?*

"Exploring the Theory of Intelligent Design"
Produced by Illustra Media
www.illustramedia.com

~ *EXPELLED: NO INTELLIGENCE ALLOWED*

Highly endorsed by actor, economist, and financial expert, Ben Stein. Don't miss it.
Produced by Premise Media Corp.

ACKNOWLEDGMENTS

I thank James Perloff, author of *Tornado in a Junkyard*, for his encouragement and advice early on.

I am grateful to Dr. J. Douglas Oliver, Professor of Biology at Liberty University, who reviewed the manuscript and provided valuable commentary.

Tom Dudley and Stacy Pfluger, Instructors of Biology at Angelina College, while not necessarily agreeing with the book in total, provided constructive criticism.

Kathy Ide critiqued an early manuscript and provided many useful suggestions for improvement.

I thank Jeanelle McCall for her contributions to the cover design.

Finally, I am grateful to my wife Jennifer, and friends John Sharp and Dan Pfluger, for reviewing the manuscript and providing feedback from the "layman's" perspective.

ENDNOTES

1. Eldra P. Solomon, Linda R. Berg, and Diana W. Martin, *Biology*, 6th ed. (New York: Saunder College Publishing, 2002), 13.

2. metaphysics. *The American Heritage Dictionary of the English Language*. 4th ed. Houghton Mifflin Company, 2004. http://dictionary.reference.com/browse/metaphysics (accessed: February 10, 2007)

3. Phillip E. Johnson, *Darwin on Trial*. 2nd ed. (Downers Grove, Illinois: Intervarsity Press, 1993), 150.

4. International Society for Complexity, Information, and Design. http://www.iscid.org/encyclopedia/Universal_Probability_Bound (July 1, 2007).

5. Jonathan Wells, *Icons of Evolution* (Washington, DC: Regnery Publishing, Inc., 2002), pp 14-19.

6. Michael Denton, *Evolution: A Theory in Crisis* (Chevy Chase, Maryland: Adler & Adler, 1986), 262.

7. Thomas Woodward, *Darwin Strikes Back: Defending the Science of Intelligent Design* (Grand Rapids, Michigan: Baker Books, 2006), chapter 9. Woodward provides interesting commentary of the current state of Abiogenesis research.

8. Roland F. Hirsch, "Darwinian Evolutionary Theory and the Life Sciences in the Twenty-First Century" in *Uncommon Dissent: Intellectuals Who Find Darwinism Unconvincing*, ed. William A. Dembski (Wilmington, Delaware: ISI Books, 2004), 230.

9. Illustra Media, DVD, *Unlocking the Mystery of Life*, USA: Illustra Media, 2002.

10. For additional information, see http://www.panspermia-theory.com

11. William A. Dembski, ed. *Uncommon Dissent: Intellectuals Who Find Darwinism Unconvincing*, (Wilmington, Delaware: ISI Books, 2004), xxxiv.

12. Richard Halvorson, "Questioning the Orthodoxy: Intelligent Design Theory Is Breaking the Scientific Monopoly of Darwinism," Harvard Political Review (May 14, 2002), as quoted in Dembski, *Uncommon Dissent*, xxxiv.

13. Henry Morris, *Men of Science; Men of God* (Green Forest, Arkansas: Master Books, 1982).

14. For a list of society fellows at The International Society for Complexity, Information, and Design (ISCIS), visit their website: http://www.iscid.org.

15. entropy. Merriam-Webster's Online Dictionary. Merriam-Webster, Inc. http://www.merriam-webster.com/dictionary/entropy (accessed: July 10, 2010).

16. P.W. Atkins, *Physical Chemistry* (San Francisco: W.H. Freeman and Company, 1978), 123.

17. Henry M. Morris and John D. Morris, *The Modern Creation Trilogy*, Vol II, Science & Creation (Green Forest, Arkansas: Master Books), 141-146.

18. George Wald, "The Origin of Life," *Scientific American* 191 (1954): 44-53.

19. Charles Darwin, *The Origin of Species*, 6th ed. (Amherst, New York: Prometheus Books, 1991), 273.

20. Solomon, *Biology*, 438.

21. Darwin, *The Origin of Species*, 273.

22. Darwin, *The Origin of Species*, chapter X, 268.

23. Darwin, *The Origin of Species*, chapter VI, 136.

24. Michael Behe, *Darwin's Black Box* (New York: Simon & Schuster, 1996), 68.

25. Behe, *Darwin's Black Box*, 69.

26. Darwin, *The Origin of Species*, chapter VI, 139.

27. Solomon et al., *Biology*, G-42.

28. Darwin, *The Origin of Species*, chapter II, 31.

29. evolution. Merriam-Webster's Medical Dictionary. Merriam-Webster, Inc. http://www.merriam-webster.com/medical/evolution (accessed: July 19, 2010).

30. The National Center for Science Education. "Talking Points for the Public Hearing of the Texas State Board of Education." http://www.texscience.org/files/ncse-talking-points.htm (accessed: March 30, 2007).

31. The National Center for Science Education. "Talking Points for the Public Hearing of the Texas State Board of Education. http://www.texscience.org/files/ncse-talking-points.htm (accessed: March 30, 2007).

32. Wells, *Icons of Evolution*, 155.

33. Johnson, *Darwin on Trial*, 25.

34. Johnson, *Darwin on Trial*, 27.

35. Jonathan Wells, *The Politically Incorrect Guide to Darwinism and Intelligent Design* (Washington D.C.: Regnery Publishing Inc.), 64.

36. Solomon et al., Biology, 438. Although I had a bit of fun with the textbook Biology, I commend the authors Eldra Solomon, Linda Berg, and Diana Martin on a well-written and beautifully illustrated textbook.

37. Wells, *Icons of Evolution*, 186.

38. Wells, *Icons of Evolution*, 182.

39. Darwin, *The Origin of Species*, 59.

40. Herbert Spencer's entire 1852 essay can be found at http://www.victorianweb.org/science/science_texts/spencer_dev_hypothesis.html.

41. Wolf-Ekkehard Lönnig, "The Evolution of the Long-Necked Giraffe (Giraffa camelopardalis L.)--What Do We Really Know?" (Part 1) http://www.weloennig.de/Giraffe.pdf (March 11, 2007).

Author's note: The selected quote is meant solely to demonstrate evolutionary gaps, and not as a reflection on the views Mr. Lönnig holds regarding Darwinism, whatever they may be.

42. Behe, *Darwin's Black Box*, 34.

43. Michael Denton, *Evolution: A Theory in Crisis*, 218-219.

44. Jonathan Sarfati, *Refuting Evolution: A Handbook for Students, Parents, and Teachers Countering the Latest Arguments for Evolution* (Green Forest, Arkansas: Master Books, 2002), 57-58.

45. Morris and Morris, *The Modern Creation Trilogy*, Vol II, 70.

46. Jonathan Wells discusses *Archaeopteryx* in Chapter 6 of *Icons of Evolution*.

47. Christopher P. Sloan. "Feathers for T. rex? New birdlike fossils are missing links in dinosaur evolution." *National Geographic* (November 1999).

48. This quote is from the first edition of *The Origin of Species*. It is interesting that this statement was altered in subsequent versions, and the hypothetical black bear-to-whale scenario downgraded from explicit to implicit. The first edition quote can be found at http://darwin-online.org.uk/content/frameset? itemID=F373&viewtype=text&pageseq=1 (May 21, 2010).

49. Jonathan Wells goes into a more detail in his book *The Politically Incorrect Guide to Darwinism and Intelligent Design*, 39-41.

50. Sunderland, *Darwin's Enigma*, 89.

51. James Perloff, *Tornado in a Junkyard* (Arlington, Mass.: Refuge Books, 1999), 90.

52. Marvin L. Lubenow, *Bones of Contention* (Grand Rapids, Michigan: Baker Books, 1992), 178-179.

53. Casey Luskin, "Human Origins and Intelligent Design," *Progress in Complexity, Information, and Design*, Vol 4.1, July 2005. Online. http://www.iscid.org/pcid/2005/4/1/luskin_human_origins.php (July 15, 2010).

54. Quote is from Berlinski's response to a letter of criticism. David Berlinski, "The Deniable Darwin" in *Uncommon Dissent: Intellectuals Who Find Darwinism Unconvincing*, ed. William A. Dembski (Wilmington, Delaware: ISI Books, 2004), 298.

55. Lubenow, *Bones of Contention*, 182.

56. Denton, *Evolution: A Theory in Crisis*, chapter 12. Also, Denton cites the following reference as a key source of data for his work.

57. Dayhoff, M.D. Atlas of Protein Sequence and Structure, National Biomedical Research Foundation (Silver Spring, Maryland, 1972), vol 5, Matrix 1, p D–8.

58. Denton, *Evolution: A Theor y in Crisis*, 281.

59. Denton, *Evolution: A Theory in Crisis*, 281.

60. Denton, *Evolution: A Theory in Crisis*, 285.

61. Darwin, *The Origin of Species*, chapter VI, 131.

62. Darwin, *The Origin of Species*, chapter X, 253.

63. Darwin, *The Origin of Species*, chapter X, 251.

64. Darwin, *The Origin of Species*, chapter X, 257.

65. Darwin, *The Origin of Species*, chapter VI, 127.

66. Darwin, *The Origin of Species*, chapter X, 262.

67. Robert C. Koons, "The Check Is in the Mail: Why Darwinism Fails to Inspire Confidence" in *Uncommon Dissent: Intellectuals Who Find Darwinism Unconvincing*, ed. William A. Dembski (Wilmington, Delaware: ISI Books, 2004), 21.

68. Phillip E. Johnson, *Defeating Darwinism* (Downers Grove, Illinois: Intervar-sity Press, 1970) 83.

69. Darwin, *The Origin of Species*, chapter VII, 167.

70. William A. Dembski, *The Design Revolution* (Downers Grove, IL: InterVarsity Press, 2004), 263.

ABOUT THE AUTHOR

Kirk Stephenson holds a Ph.D. in chemistry; worked in the specialty chemical industry for twenty-four years, in both research and management positions; served as an instructor of chemistry; is the inventor/co-inventor on eleven patents; and is licensed as a patent agent (United States Patent and Trademark Office).